UNCONFINED VAPOUR CLOUD EXPLOSIONS

A comprehensive and critical study of incidents, experimental work and theory with the object of highlighting and suggesting possible solutions to the problems of prediction, prevention and protection in the light of the present state of experience and knowledge

UNCONFINED VAPOUR CLOUD EXPLOSIONS

KEITH GUGAN
PhD, CEng, FIChemE
Research Fellow appointed by
The Institution of Chemical Engineers

THE INSTITUTION OF CHEMICAL ENGINEERS
in association with
GEORGE GODWIN LIMITED

Published in 1979 by
The Institution of Chemical Engineers,
George E. Davis Building,
165–171 Railway Terrace,
Rugby, Warks, CV21 3HQ

in association with
George Godwin Ltd,
Builder House,
1-3 Pemberton Row,
Red Lion Court,
Fleet Street,
London EC4P 4HL

(ISBN 0 85295 114 0 and 0 7114 5102 8)

Printed by INPRINT of Luton (Designers & Printers) Ltd.

UNCONFINED VAPOR CLOUD EXPLOSIONS

by Dr Keith Gugan
CEng, FIChemE

CONTENTS

LIST OF LINE DRAWINGS

LIST OF LINE DRAWINGS (Contd.)

LIST OF PHOTOGRAPHS

(The photographs are contained in a 16-page section located at the centre of the book)

LIST OF PHOTOGRAPHS (Contd.)

FOREWORD

by Professor D.C. Freshwater

I am happy to be invited to write the foreword for this highly significant publication. I cannot pretend to be independent, since I was Chairman of the Steering Committee of the IChemE which set Dr Gugan on his way as Institution Research Fellow. Having said that, it must be made clear that the work is Dr Gugan's and not that of a Committee of which he was rapporteur. All we did was to suggest guidelines, to help with contacts and to comment on the various drafts he produced. He was charged with making a thorough and critical survey of the problem of the Unconfined Vapour Cloud Explosion. The present publication is evidence of how well he succeeded in his appointed task.

Such is the timeliness of this topic that no apologia is needed for this book. Even were this not so, this work would stand in its own right as a substantial contribution to the chemical engineering literature. There is no comparable publication which collects and analyses the data of all known incidents up to early 1977. The novel yet simple theoretical treatment offers a real way forward to prediction of damage effects due to explosions of unconfined vapour clouds. Yet this is only a beginning – as Dr Gugan himself acknowledges. Much remains to be explored not only in the area of theoretical prediction but in the ancillary areas of prevention and protection. A whole new subject field is being opened in which chemical engineers will be amongst the leading pioneers. And they will start with this book as their guide. I have no doubt that it will quickly become a standard reference and, although in a rapidly growing topic it will undoubtedly be superseded by new developments, nevertheless it will remain as a landmark. On behalf of the Steering Committee I congratulate Dr Gugan on his work.

Finally may I add a word of praise for the foresight and venture of the Institution of Chemical Engineers and particularly the Research Committee and Publications Committee on their decision to publish this report in its entirety. At a time conspicuous in this country for the timidity of public bodies it is encouraging to see such enterprise. I am sure that it will prove to be a well founded judgment.

D.C. Freshwater
Loughborough
September 1978

Acknowledgement

I gratefully acknowledge the active help and encouragement of my Steering Commitee and thank them for their lively interest in the content and progress of this Fellowship. I also acknowledge the stimulation provided by, and the patient restraint of, the Research Committee of the Institution of Chemical Engineers. Not least, I thank all of those who gladly gave their time to discuss and correspond with me on this difficult subject, particularly those in the Processing Industry who contributed their experience.

I commenced my work on this Fellowship on 3 August 1977. That it has come to this state at all has been due to the patient forbearance of my family, and the great and unremitting efforts of my secretary, Mrs Teresa Dickerson.

K. Gugan

I
PREAMBLE

"On completion of present third-generation fuel/air explosive development work, with a 500 kg methane charge it will be possible to obtain an excess pressure* of 0.9 kg/cm^2 (major damage, sinking) at distances of 100 to 130 m from the blast boundary*" – Gerry Johannsohn, Bremen, *International Defence Review,* (1976) 6, 992-996

This statement indicates what is currently the realistic expectation of the performance of fuel/air explosive devices. It is based upon the actual performance of existing weapons. The article goes on: "The measurable excess pressure peak of methane is only (*sic*) 22 to 24 kg/cm^2 and, by comparison to ethylene oxide, methane is a slow reacting gas."

Before putting this into perspective with the problems which face engineers, planners and others, whose daily lives bring them into contact with the unconfined vapour cloud explosion risk–however remote in real terms it may be–it may be useful to add further dimensions to these statements. Methane is one of the least reactive hydrocarbons in combustion terms. It has a low burning velocity and a high ignition temperature. 500 kg of methane will, under normal conditions of temperature and pressure (NTP), form a stoichiometric mixture with air of 7000 m^3 – a volume equivalent to a sphere of diameter 23.7 m or a hemisphere of diameter 29.9 m. In these cases the burnt cloud boundary (the blast boundary?) would be approximately 25 m and 30 m respectively from the centre of symmetry. The maximum over-pressures found at the centre of the blast cloud are 21 to 23 atm and at 125 m to 160 m from the centre 0.9 atm above normal atmospheric pressure.

Of course, the above refers to weapons and these are contrived to maximise their destructive potential. The effects quoted are achieved by detonators spaced throughout the combustible volume. No reliance is placed upon spontaneous transition to detonation. Nevertheless, the effects represent what may be presently taken to be the upper bound of foreseeable potential combustion behaviour. Although fuel/air explosives largely comprise hydrocarbons and olefin oxides, one non-military example of clearly analogous combustion behaviour is known to me[1], which involves the release into the air of a high-pressure mixture of hydrogen and carbon monoxide (principally) at a temperature greatly in excess of its spontaneous ignition value. This will be referred to in more detail later.

* These and other specialist terms are defined in the Glossary.

The problem of unconfined vapour cloud explosions (UVCEs) has only relatively recently been recognised for what it is—one of the few remaining threats of total devastation to process plants and their surroundings. No textbook on combustion science published prior to 1970[2-11] and few published after 1970[12] make any reference to the possibility of "normal" combustion behaviour in the open atmosphere—on whatever scale—being accompanied by damaging pressure effects. Manuals of chemical engineering and courses of study which involve design or safety considerations (for example, reference[13]) ignore this latent hazard. Although learned papers were published as early as 1946 dealing with the fluid mechanics of atmospheric disturbances of a generally similar though artificial type[14], the crucial link between fundamental flame behaviour and the fluid mechanics was not forged.

This is not to suggest that this was a trivial oversight. As will become clear, theoretical and conceptual difficulties remain. But, more importantly, the incentive to undertake the necessary work was absent until unequivocal experience of the phenomenon was gained. Furthermore, as appeared to be self-evident and as experimental work has since suggested, the scale of testing necessary to simulate real behaviour in unconfined flammable volumes is inevitably vast.

Controversy still exists, even amongst the experts, over the correct description of the combustion process involved (deflagration or detonation) or indeed whether a third, as yet unnamed, category of gas-dynamic event occurs. Deflagration and detonation are reasonably well understood combustion mechanisms and have been studied extensively in the laboratory, although mainly in unidimensional or uniaxial geometries. The principal differences in the unconfined situation are that flame movement is in three dimensions and, for blast effects to result, inertial considerations may well be important.

There is even dispute over the description of the phenomenon. Professor Rasbash[15] has pointed out that in no case known to him have blast effects resulted with total absence of "confinement"; chemical plant structures, vegetation or conventional buildings have been involved. Indeed, the only two genuinely unconfined cases disclosed by the preparatory studies for this report both involved initially highly turbulent gases—which effect is, of course, created by the types of "confinement" normally found in vapour cloud explosion situations.

The use of the word "vapour" may also be questioned. It is, perhaps, evoking a distinction without a difference to state that the description "vapour" must include those compounds which exist as permanent gases at normal temperatures and pressures. However, vapour in this context must also include fogs, mists and sprays of flammable liquids. Whether there is any lower limit to the vapour pressure which flammable or merely combustible liquids must exhibit or even whether explosive effects could be created by a cloud of combustible dust is not known.

So long as these qualifications are borne in mind and the words "unconfined vapour cloud explosion" are not given an unduly restricted interpretation, they provide an accepted and, one hopes, a generally acceptable description of the phenomenon.

II
INTRODUCTION

The purpose of this report is to provide those interested in the phenomenon with an up-to-date résumé of relevant experience and a review of current thinking as reflected in technical publications, discussion at recent meetings and correspondence. To this end I have requested from every major petrochemical company and chemical plant contractor in the world details of incidents or rationales behind their hazard appraisals. Consultants and University Departments have also been contacted on a similar basis. What follows is hoped to be an authoritative dissertation on the basis of the most recent information.

It was impossible to perform experimental work in the course of this Fellowship. Nevertheless, whilst preparing this document a number of ideas for such work have evolved, which are set out herein.

Many who may have occasion to refer to this report will have neither combustion experience nor the appropriate technical background. Accordingly, the body of the text is simplified and large tracts of theory have been consigned to the Appendix. Also, the more fundamental concepts are explained with the object of enhancing understanding. This will mean that those who have the necessary skills or experience may find parts set out in what they would consider elementary terms. I crave their indulgence. My approach has been essentially pragmatic: rigorous theory may, one day, enable a complete appraisal of the hazard—but that has not happened yet. The essence of this study is to quantify and correlate damage potential and actual damage—as revealed by incidents—and further to compare this with mortality and damage statistics from incidents involving more conventional explosives. Such an approach, in my view, avoids the invidious arguments of conventional risk analysis because it does not seek to compare the relative risks of injury or death arising from voluntary, as distinct from imposed, conditions of living.

Accordingly, the foundations of this report must be the incidents themselves. Analysis of each incident could, in principle, enable an accurate assessment of the damage to be made and the whole would then form an edifice of experience from which the damage potential of various stored or process quantities of different materials could be determined. As will appear later, this objective is at the present time unfortunately not capable of achievement to any really satisfactory extent.

The scope of this study is broadly as set out in the Table of Contents. Apart from the *quasi*-statistical aspect alluded to above, the intention is to undertake a detailed review of a small number of well documented cases, as these are not only object lessons in the effects of UVCEs, but also they indicate the usefulness of and the care which must be taken in applying conventional diagnostic techniques. It is proposed that the conventional techniques of analysis should be extended and a number of "new" methods are suggested.

The theoretical aspects are dealt with in two stages. The section on *Combustion and Physical Fundamentals* is a simplified presentation of well-established concepts wholly confirmed by experimental studies. Later in the report, however, certain physical aspects will be dealt with which may be novel—certainly most of which are unproven. Here the technicalities are more within the field of the specialised reader. This section will be closely coupled to the sections on *Well Documented UVCEs* and *Experimental Work.*

It is intended also to present a picture of the present state of knowledge regarding the UVCE risk in respect of the inherent properties of materials and their storage or process volumes, pressures and temperatures. In relation to the risk as it is currently assessed aspects of prevention and protection will be highlighted.

Finally, the report will deal with possible advances in the understanding of the phenomenon and in the elimination, or at least amelioration, of its effects.

In this review "causation" is not a live issue except insofar as it is relevant to the precise mode of primary emission of the flammable material into the free atmosphere. It is recognised that how the emissions came into being is not known in some instances[16–18] and in dispute in others[19–21]. Indeed, in some cases even the nature of primary emission is in doubt[22]. ("Primary" is here used to mean principal or predominant, *ie*, in magnitude, but in most cases it is also synonymous with first chronologically.)

In the time available I have not been able in most cases, either to check damage patterns or, by discussion with investigators or first-hand perusal of evidence, assessments of damage in terms of, for example, TNT equivalents. Largely, assessments have been accepted without question, but it must be remembered that few incidents have been studied by those familiar with blast or TNT equivalents. There will undoubtedly be errors of an indeterminate magnitude arising from this circumstance.

III
COMBUSTION AND PHYSICAL FUNDAMENTALS

General

It is not intended to present here more than a brief and elementary review of some of the concepts which may arise later. Most technical readers will aready be aware of the matters dealt with here which are to be found in most texts on combustion[2-11]. A pertinent text has recently appeared in the American Institution of Chemical Engineers Monograph Series[23].

It should be appreciated that the primary concern in this context is with homogeneous combustion where both the oxidiser and the "fuel" are gaseous. Two situations may be distinguished—one in which the oxidiser and fuel are premixed and the other where they are not.

Additionally, one important case of "heterogeneous" combustion must be considered and that is the combustion of a mist. Here, whilst the fuel (the vapour derived from the droplets) and the oxidiser are not premixed, the character of the system induces flame behaviour similar to that in premixed gaseous systems.

All the flame systems with which we are concerned create high temperatures. These high temperatures in turn create products with high specific volumes and thus generate expansion forces (in addition to the obvious buoyancy forces). Most saturated hydrocarbons, when they burn, form products of a higher intrinsic volume than that of the reactants, but this volume increase is trivial compared to that caused by the high temperatures. The temperature increase on combustion is a function principally of the heat of combustion of the "fuel" and the nearness to perfect equivalence of fuel and oxidiser in the mixture. Perfect equivalence of fuel and oxidiser occurs when both are present in "stoichiometric" proportions. Other factors which affect the temperatures attained are whether or not the burning gases can expand and the rate of heat loss.

It is important to recognise that these remarks are appropriate to overall combustion characteristics in the particular system considered, but regions within that system may behave quite differently. In other words, the behaviour described above is average behaviour. Departures arise when rates of combustion become enhanced: small volumes within the system display significantly higher temperatures and pressures. Because the small volumes concerned are coincident with the propagating flame, they can and do exert their effects at all points (but not simultaneously) to which the flame propagates.

We have introduced somewhat incidentally the basic energy releasing phenomenon:

the flame. The flame is, of course, fundamental to practically all combustion systems. Henceforward we shall consider "fuels" of the hydrocarbon type, unless otherwise specified, and, as oxidiser, air. Ordinary flames are somewhat surprising entities. They are, naturally, hot, but Weinberg[24], to my knowledge, was the first to show that they exhibit accelerations of 10^4 *g* and above.

Combustion of Gaseous Mixtures

Not all mixtures of fuel vapours and air will burn. To sustain a flame mixture compositions have to be within the flammable range. This range, which varies according to the flammable material, is bounded by the lower and upper flammability limits. These limits are usually defined in volumetric or molar terms: fuel vapour volume as a percentage of the total vapour/air mixture volume. The flammable ranges of all materials naturally encompass the appropriate stoichiometric proportions.

Flammability limits are usually measured using an apparatus and a technique which have become standard[25]. It is clear[4], however, that no fundamental significance is to be attached to limits which are measured in this way; indeed flame propagation in the standard apparatus may be achieved between wider flammability limits if the technique employed is non-standard. The proximity of surfaces to the flame has a severe effect, especially near the limiting condition for propagation, and this is exemplified by the observation of wider limits if apparatus larger or different than the standard is employed[26,27].

Whatever the absolute significance of flammability limits, they do exist. Moreover, comparative limits are useful and in most instances the range of flammable concentrations is correctly circumscribed by experimentally determined limits. Flammability limits are temperature- and pressure-dependent.

The effects, which occur when a flame travels through a mixture of flammable vapour and air, depend upon such factors as the initial pressure and temperature of the mixture and its precise composition as well as turbulence, homogeneity and degree of confinement. For most hydrocarbon/air mixtures maximum rates of flame propagation occur near the stoichiometric composition and generally are enhanced by increased pressure, temperature and turbulence. This is universally true in respect of temperature and turbulence, but there are exceptions so far as increased pressure is concerned; methane, for example, is retrograde in displaying a progressive lowering of burning velocity with increasing pressure.

Rates of flame propagation are characterised by the "burning velocity". This is the concept of a steady velocity of propagation of a self-similar flame perpendicular to itself into unburned gases. Flame propagation in gases exhibiting streamline flow at low Reynolds numbers proceeds at the rate determined by the "laminar burning velocity". Generally, tabulated burning velocities are those measured under laminar or static flow conditions and represent the lowest possible inherent flame propagation rates. Typically these range from about 0.37 m s^{-1} for methane, through 0.64 m s^{-1} for ethylene, 1.0 m s^{-1} for ethylene oxide and 1.3 m s^{-1} for acetylene, to 2.9 m s^{-1} for hydrogen. The saturated hydrocarbons as a group manifest burning velocities in the range 0.35 to 0.45 m s^{-1}. The unsaturated hydrocarbons display somewhat higher burning velocities, although, with the exceptions of the first members of their homologous series, they do not lie outside the range 0.4 to 0.5 m s^{-1}.

Turbulent burning velocities have been measured and are reported in the literature[28]. Because turbulence is, in principle, without limits, there is no single turbulent burning velocity; each turbulent burning velocity is matched to the degree of turbulence extant. The "degree" of turbulence is quantitatively described by the size of the Reynolds number (*Re*), but for any given value of *Re* the flow mass-rate and system characteristic dimension are inversely related, thus two further factors have to be introduced to fully describe (both quantitatively and qualitatively) flow behaviour: *intensity* of turbulence–related to mass-rate, and *scale* of turbulence–related to the characteristic dimensions of the flow system. Of these factors intensity would appear to be the more significant influence on flame propagation. Turbulence typically enhances burning velocities by a factor of 10.

Deflagration

Burning velocity is not the same as flame speed. In most fire or explosion situations the burnt and unburnt gases are moving; the burning velocity is supplemented by the translational velocity of the bulk motion.

Flame temperatures depend upon the nature of the flame and the pressure, temperature, composition and flammable constituent involved. Most hydrocarbons at or near stoichiometric proportion in air and burning normally at NTP create temperatures in the range 1800–2000°C. This temperature rise leads either to a final pressure (in a system affording total confinement) of 7 to 7.5 atm or to a final volume (in an isobaric system) of 7 to 7.5 times the original volume; both "final" conditions representing equilibrium without heat loss.

As 7- to 7.5-fold is close to 8-fold, the latter will be the "expansion" factor adopted henceforth. Consider a flammable mixture in a tube with one end closed; if ignition occurs at the closed end, the resulting flame will exhibit (in an isobaric situation) a flame velocity of v and a bulk velocity, due to expansion, of $8v$. The total flame speed relative to a stationary reference point would therefore be $9v$. Spherical flame propagation outwards from a central ignition source would generate, in similar circumstances, a flame speed of $3v$. Clearly, flame speeds can become high, especially in the tube situation where the displacement of gas ahead of the flame creates pipe-flow turbulence, which further enhances the fundamental rate-controlling phenomenon: burning velocity. Flame speed without turbulence in such a system would be about 24 m s^{-1} for hydrogen, but turbulence could lead in principle to "normal" flame speeds of 240 m s^{-1} or more. Similarly for propane the "normal" flame speed could reach 90 m s^{-1}.

These figures are based on published data, but are not strictly comparable with each other and are, in any event, related to constant pressure behaviour. This conception is convenient, but imaginary. In the unidimensional behaviour cited for the flame travelling in the tube there are pressures generated by pipe frictional effects. Additonally, gases, in common with all matter, have mass and therefore exhibit inertial effects; because gases are compressible, these show up transitorily as pressure. A further pressure effect arises from the change in momentum across the flame itself. Which effects predominate is not particularly relevant here; all are ultimately related to burning velocity in various ways.

Although unidimensional behaviour is not appropriate to this study, it is worth pursuing further. Two factors which it shares with three-dimensional behaviour are inertial and momentum-generated pressure effects. These do not attain the same magnitudes (after due allowance for pressures induced by flow) within the same time scales, however, because of the three-fold reduction in expansion rate conferred on the three-dimensional system by the geometry. Moreover, the auto-accelerating effect of pressure on those gases whose burning velocities are positively pressure dependent is also substantially reduced, at least within comparable initial distance and time scales.

Thus, two factors, in principle, render spherically symmetric systems radically different from unidimensional ones—the absence of an interrelating solid surface and the geometric "dilution" of the effect of expansion on flame propagation. Are there compensating features? In practice, of course, most vapour clouds burn close to the ground. But it is unlikely that this would create much turbulence or that what was generated could affect the cloud in depth. Again, in practice, most vapour clouds contain rigid or semi-rigid objects, the effects of which will depend upon their size (relative to the cloud), character and density of distribution. But is there an *intrinsic* feature of spherical geometry which assists flame propagation?

There is one, I would suggest, the effects of which are more apparent if we return to the isobaric concept. The effects are present if the system is considered to behave naturally, but they may be the more readily visualised with the help of this artifice. Spherical elements of flammable mixtures outside the flame (assumed to be propagating radially from a central ignition source) move away from the centre of symmetry before they are consumed by the flame, at rates and to extents appropriate respectively to the flame speed and their initial distances from the centre.

Because of the isobaric constraint, volume is conserved, but, because of their outward movement, spherical elements display greater area—consequently their thickness must decrease inversely as the square of the ratio of their radial position after displacement to their initial position. A spherical element of unit thickness at unit distance from the centre will exhibit one quarter of its original thickness when displaced to two units of distance from the centre of symmetry. The matter in this element will be redistributed circumferentially (analogously to spreading butter on bread). Whether this redistribution, or spreading, occurs turbulently or non-turbulently depends upon the rate of displacement. If the process is turbulent, it is not possible to say which of the elements of turbulence—scale or intensity— predominates. It would seem more likely that redistribution is microscale and thus of enhanced intensity.

Such a conception means that all vapour clouds may be expected to be capable of creating and supporting accelerating flames. Some freely expanding spherical flames even on the laboratory scale display surface roughness when burning non-turbulent gases[5].

The above considers in qualitative terms the nature of flames without indicating in depth the role they have in producing pressure effects. Explosions are phenomena where the pressure effects created are usually distinctly audible. The effects on the ear are within the frequency and amplitude response of that organ. Most explosions are well above the minimum audible value. Our concern is with explosions in which either the frequency or amplitude, or both, of the pressure wave is capable of doing damage well beyond the confines of the explosion cloud.

Commonly explosions arise in situations of near total initial confinement. They occur and can be explained without resort to "abnormal" flame behaviour. In a situation where the space involved is not unduly elongated (no dimension exceeds say three times—or less than one-third of—the square root of the product of the other two), a flame burning a flammable mixture distributed throughout the volume will by normal flame propagation cause an increasing pressure to be felt on the boundaries of that volume due to the expansion effects of the hot products of combustion. The magnitude of the pressure created at any instant will *approximately* correspond to the ratio of the sum of the volumes of unburnt and burnt gas (the latter evaluated at the initial pressure) to the original volume, multiplied by the original pressure of the system. The word "approximately" is used because the pressure effect created is not quite synchronous with the appropriate condition in the combustion process. The lack of exact synchronisation arises because the pressure "information" is conveyed to the boundaries of space at a finite velocity: the velocity of sound. In normal aerodynamic or pneumatic systems pressure information cannot be transmitted at a greater speed.

The degree of pressure lag in any normal system depends on the scale of the system and the ratio of the flame propagation velocity to the velocity of sound in the *unburnt* gas. Large spaces and fast flames give actual instantaneous boundary pressures well above and at the same time well below instantaneous average values, though not at the same locations.

Most confined explosions occur in relatively small volumes compared to UVCEs or in relatively small pockets of flammable gases in large volumes. The explosive effects are created when a boundary fails under the developing pressure. Debris may be projected moderate distances and blast damage may result. The creation of a pressure "discontinuity" or shock wave which does damage causes no surprise—this discontinuity arises at the boundary of the volume and is held there until failure occurs. When the boundary does fail, a "step-form" of pressure wave is released having an amplitude above the surrounding pressure initially equal to the failure over-pressure. This wave propagates at a characteristic velocity close to that of sound in the ambient atmosphere and dissipates with distance. It conveys energy and does work (causes damage) as it travels.

The pressure or shock wave formed in this way is a genuine "discontinuity" in that within a few molecular mean free paths the pressure rises from ambient values to that of the shock wave. The discontinuity, once formed, persists until the pressure amplitude has reached fairly low, acoustic, levels. Quite where this happens, *ie*, at what distance from the explosion centre, in any particular case depends upon the initial over-pressure.

Hitherto the consideration has been of either ordinary flame processes in unconfined or partially confined (open-ended tube) systems—not, as yet, associated with pressure discontinuities, or in initially closed systems—normally associated with pressure discontinuity. The emphasis here is that both categories of event have been considered to involve normal flame propagation. Normal propagation is termed *deflagration* and is to be distinguished from "abnormal" or unusual flame behaviour by its rate of flame propagation being less than the velocity of sound in the unburnt gases.

Detonation

At the other extreme of flame combustion behaviour is *detonation*—an "abnormal" or unusual mode of flame propagation in that no combustion device as yet makes use of the phenomenon and few accidental occurrences would seem to have involved it. Detonation, in sharp contrast to deflagration, is a combustion process which propagates at many times the velocity of sound in the unburnt gases—to be specific, at the velocity of sound in the hot products of combustion. Stable detonation is an accurately predictable phenomenon, susceptible of experimental verification. The theory is established, although there have been recent changes in thinking on the detailed mechanics of the process[29].

Briefly, detonation is a self-sustaining physico-chemical event. The passage of a strong shock wave through some combustible mixtures causes their instantaneous compression, heating, spontaneous ignition and reaction. The energy released by reaction is fed into the shock wave and sustains it. Compression yields a high potential reaction energy density—prior to reaction—and the high temperature ensures that this energy or a substantial part of it is released rapidly. A frequently cited example is that of 20% hydrogen in air[2,23,30]. In this case the shock wave propagates at 1600 m s^{-1} and causes an initial pressure and temperature rise, before any reaction occurs, to 23 atm and 1350 K. Reaction is immediately initiated and becomes virtually complete within a few tens of microseconds[31], by which time pressures have fallen to some 13 atm although the temperature will have risen beyond 2400 K. The whole process is complete within a space of about 50 mm behind the shock front.

Not all flammable mixtures will stably propagate detonations. Generally most will within their ranges of flammability but not close to their limits. Some "fuel"/air mixtures in particular are sensitive to the geometry of propagation[32,33]. Indeed, some "fuel"/oxygen mixtures, *eg*, equimolar acetylene/oxygen, do not detonate in three-dimensional symmetry unless the power of the ignition source exceeds a minimum critical value[29].

Reverting to the main theme, in "fuel"/air mixtures much attention in recent years has been given to the transition from deflagration to detonation. This is not simply achieved by flame acceleration, although this is the primary driving force. The spontaneous transition has been studied in tubes but has not been experimentally observed in three-dimensional geometry[34–37]. Where detonation has occurred in three dimensional systems, powerful initiators, *eg*, explosive detonators, have been used. In tubes it appears that enhancement of pressure beyond a certain value ahead of the travelling flame front causes spontaneous ignition of one or more pockets of flammable gas[29], which then emit strong supplementary pressure waves. These pressure waves ultimately coalesce with spontaneous ignition of the flammable mixture, whereupon the original combustion process is left behind by the rapidly accelerating and closely coupled detonation. The original flame dies away when it encounters only combustion products from the detonation.

To date theoretical explanations have failed to explain why, if spontaneous transition to detonation in three dimensions does not or is unlikely to happen, flammable clouds manifest blast. In the Appendix a more detailed review of the theories will be presented with contemporary thinking on the subject.

Fire-balls

The foregoing relates to flame spread over or through flammable gases. It is as relevant to fire-balls as UVCEs, although in the former the volume of truly flammable gas/air mixture (within the limits of flammability) forms a relatively thin shell enveloping over-rich mixtures. When flame has spread to envelop an over-rich cloud, combustion of the "fuel" in depth occurs. This proceeds relatively slowly compared even to ordinary rates of premixed combustion. The flame exists at the boundary between fuel and air and is sustained by fuel and air diffusing towards it and products of combustion diffusing away. The diffusion alluded to is not conventional molecular diffusion, although this is undoubtedly involved, but turbulent eddy diffusion, either present initially or developing as a result of buoyancy effects. Buoyancy and large-scale turbulence are evident in fire-balls, all of which rise from the ground with the characteristic mushroom effect.

Heterogeneous Systems

These include inadequately mixed *homogeneous* or single-phase gas systems as well as two-phase, solid- or liquid-in-air systems. The discussion here is restricted to the combustion behaviour of liquid-in-air systems. More details are to be found in the literature[2,4,9,12].

The combustion behaviour of liquids depends upon their degree of dispersion—that is, the specific surface they exhibit—as well as usual kinetic, thermochemical and thermodynamic considerations. Inherently, the combustion rate is low as the process is diffusion controlled. Depending upon the density and size of the droplets in suspension, the flames which surround them may be individual to each droplet or merged. Given a suspension of the right particle size even the most involatile but combustible materials will burn. The time taken for individual droplets to burn is almost exclusively a function only of droplet diameter and is unrelated to the actual material burning. Very small droplets, less than 10 micron, behave like vapour—they do not maintain their own flame envelopes. Above 40 microns they have their own discrete flames[38].

The observed burning velocities of aerosols are low, compared to premixed hydrocarbon flames. The highest values are exhibited by the coarser dispersions. Limits of flammability are somewhat more indeterminate than for gas mixtures. The less volatile materials have high upper limit concentrations. Involatile mists will burn when very rich although naturally they will not burn completely. Coarser dispersions have lower lean limit concentrations. As with homogeneous mixtures the expansion effects which result are controlled by the temperature which is attained after combustion. This is not necessarily dependent purely on thermochemical and concentration values, however, as part of the heat may be absorbed as latent heat to evaporate the residual droplets when all the oxygen has been consumed.

Although on the microscale combustion is a straightforward diffusion-based process of low intrinsic rate, this is only strictly true for droplets having a negligible velocity relative to the supporting air. Turbulent conditions within the supporting medium lead to higher rates of combustion.

In a vapour cloud formed from, for example, a flashing liquid escape droplets of all sizes would be found. Their initial size range would depend primarily on the shear they experience on emerging to the atmosphere, that is, the force largely due to the pressure within the plant. To what extent the droplets reduce in size by subsequent evaporation then depends upon the atmospheric temperature, the vapour pressure characteristics of the material concerned and the concentration of vapour already achieved. Indeed, recondensation of vapour produced by the initial flash can occur. Clearly, conditions within a flashing escape are non-uniform and the course of combustion in such a medium would be anticipated to be largely unpredictable.

Combustion in heterogeneous systems can in principle give rise to pressure effects because of the inherent gas dynamic properties of such systems. These will be considered later.

The Behaviour of the Atmosphere

Under normal conditions the atmosphere is a continuous medium exhibiting no sudden variation of density, composition or temperature. [*Sudden*, in this context, means within the space of a few molecular mean free paths (say 10^{-3} mm), or within a few microseconds.] Of course, density (or pressure), composition and temperature variations do occur—these being largely responsible for the weather. The most obvious and ever present manifestation of these variables is wind.

In elementary terms wind results from a pressure difference; the pressure difference is the driving force—the wind is the effect. However, as pressure discontinuities are unusual in the normal atmosphere, "pressure difference" means pressure gradient, that is, pressure varying more or less uniformly with distance. As pressure is a force, it causes the mass of air on which it acts to accelerate. The acceleration is limited, however, by the drag force exerted by the ground and objects upon it. The drag force, which acts in opposition to the pressure force, is a positive function of the wind velocity and therefore increases in magnitude until it just matches the pressure force. Under these conditions the wind achieves a stable velocity.

Wind is movement of air and constitutes a redisposition of mass within the atmosphere. As regions of higher pressure generally contain more mass per unit volume, the mass conveyed by the wind tends to iron out pressure gradients. The effects are not quite as straightforward as this might suggest, but the above is adequate for the purposes of this report. Approximately 0.1 Pascal/kilometre (0.0001 N/m^3) will induce a wind speed of 3 m s^{-1} in the open atmosphere.

Normal atmospheric behaviour is well described by considering the air to act as an incompressible fluid. The obvious exception to the above is the ability of the atmosphere to transmit sound waves, although even these do not produce atmospheric variations of sufficient rapidity to fall within the definition of "sudden" above. Sound is characterised by *frequency* (pitch) and *amplitude* (loudness). Frequencies from one Hertz (Hz) through 50 kHz encompass what most animals can generate or hear. As sound propagates, again in the ordinary atmosphere at normal temperatures, at about 330 m s^{-1}, the respective wavelengths are 330 m and 6.6 mm. (The velocity of sound is proportional to the square root of the absolute temperature.)

Loudness of sound is characterised by the amplitude of the disturbance, defined as half the magnitude of the difference between the pressure peaks and troughs in the acoustic field. The amplitudes of ordinary sound waves are relatively small, ranging from 20 microPascals (2×10^{-5} N/m^2) at the threshold of hearing to 63 Pascals (63 N/m^2 or 6.2×10^{-4} atm) at the threshold of pain (at 1000 Hz).

The atmosphere transmits acoustic waves to beyond the upper pressure level (63 Pa) without distortion of the typically sinusoidal (for single frequencies) wave forms. At amplitudes in excess of a certain critical level, however, irreversible changes occur during propagation and the form of the pressure wave changes with time (and distance). Those regions of the pressure wave which are at higher pressure (and thus higher temperature because of the adiabatic compression) travel slightly faster than lower-pressure, cooler regions. Consequently the higher-pressure regions of the same wave catch up with the lower-pressure regions. The converse happens at the "back" of the wave. This process generates a saw-tooth wave form with a near instantaneous pressure rise, followed by a region of falling pressure travelling behind with the same velocity. This pressure distribution and wave form are typical of a shock wave. Similar behaviour would normally occur in the case of the single pressure wave such as might be generated by an UVCE.

Degeneration of the shock waves to acoustic waves occurs by the reverse process when the amplitude of the shock has fallen sufficiently. This happens naturally as the wave energy dissipates with distance in accordance with conventional three-dimensional behaviour. The properties of shock waves are fully described in published texts[39,40,41].

All conventional explosives produce shock waves, the characteristics of which, because of their military uses, have been thoroughly studied. These are, of course, to be distinguished from gas-phase detonation shock waves, which generally have much lower pressure amplitudes. They are also different in that they are not closely coupled to the chemical reaction except for the first few microseconds of movement. Because of the severely damaging effects explosive blasts can produce, accurate correlations are now available for explosives of all types. The principal standard against which other explosives are matched and with which UVCEs are at present conventionally compared, is that of the high explosive, tri-nitrotoluene (TNT). A comprehensive collection of correlations of blast data from TNT detonations has been compiled[42] which would be difficult to find improved upon in any single paper.

One natural event, which occurs frequently and gives rise to shock waves, is lightning. This suddenly releases approximately 10^5 joule/m[42a,42b] into the atmosphere, which propagates the energy radially away from the lightning stroke first by shock wave and then by acoustic wave. The peak side-on shock pressure is inversely related to the distance from the lightning strike in such a way that at 0.02 m the shock wave overpressure is about 50 MPa and at 0.5 m it is 1 MPa.

Relatively close to a lightning strike the wave is a single shock accompanied by a wide frequency spectrum of much lower amplitude sound waves. Further away matters become complicated by the combination and interference of sound waves from other portions of the lightning channel coupled with reflection and refraction effects, producing an extremely complicated frequency pattern. The complexity arises out of the rate of deposition of energy (extremely high—within 0.1 μs) and would not be expected to occur with a lower rate of energy deposition.

The theoretical view taken in this report is different from that propounded by other workers in this field hitherto and does not assume the precise form of energy release—although for convenience hemispherical symmetry is assumed. The "ordinary" properties of the atmosphere outlined above are also assumed to apply. Details of the approach are set out in the Appendix.

Basically the energy release and energy transfer processes are separated, the former generating a large-amplitude sinusoidal pressure wave with its peak at the centre of symmetry—the pressure falling away uniformly (sinusoidally) to atmospheric at the boundary defined by the distance that a sound wave would have travelled in the time of energy release. Thereafter all parts of the pressure wave are assumed to propagate at the velocity of sound radially and symmetrically outwards in three dimensions. (A two-dimensional analogue is provided by the waves generated when a stone is dropped into water—although here only the first and principal wave is considered.) This gives a sinusoidal wave of *constant* pulse duration (equal to the *diameter* of the pressure-wave-affected zone at the completion of rapid energy release divided by the velocity of sound) but of uniformly varying amplitude—as dictated by the requirement for energy conservation.

As represented in the Appendix, it is not suggested that this approach, or the theory upon which it is based, is necessarily precise or rigorous. In my submission theoretical precision, whilst being the alternative and desirable goal, is not essential—it must, however, be consistent. The theory should be judged only upon its powers of correlation and prediction and its general compatibility with events. The theory is used to reinterpret some of the evidence presented in Section V—to, I would submit, convincing effect. Moreover, the predictions of the theory go some way to explaining flame acceleration effects, although there the theory is applied to a non-uniform and non-symmetrical system. One important factor is that those near to a UVCE apparently do not hear a bang, although they suffer violent pressure effects. The theory enables calculation of the distance, at which an audible bang would occur, to be predicted.

The theory is, of course, idealised—it assumes a uniform rate of energy release (E) and a uniform central rate of pressure rise ($\dot{p}_c$). It correlates with damage, which not only requires a certain minimum pressure to be achieved, but also a minimum rate of pressure rise. It explains the severe buffeting, which has been described by witnesses, when pressure gradients in excess of 30 kN/m^3 are generated (*cf* 0.0001 N/m^3 for a uniform wind of 3 m s^{-1})—and the severe if short-lived hurricane winds which develop.

To a considerable extent the theory is a compromise; its formulation has required the assumption of various ideal conditions. Most, however, of the important predictions of the theory—which are essentially insights into the mechanistics—are obtained only if the theory is applied to a non-ideal geometrical system.

IV
REVIEW OF ALL MAJOR INCIDENTS

This heading is ambitious and to some extent the objective is incapable of realisation. The "incidents" considered are those in which flammable gases or vapours were released into the atmosphere and ignition occurred. There have been many releases of very major proportions in which ignition did not occur but, as these have no diagnostic value in this context, they will not be considered here.

Generally, two extremes of combustion behaviour are possible following ignition of an unconfined flammable volume: the fire-ball and the UVCE. Elements of both are probably present in all cases. In fire-balls combustion energy is released almost exclusively as thermal energy; in the UVCE part of the combustion energy manifests itself as kinetic energy. Fire-balls arise where the conditions of release to atmosphere inhibit mixing between the flammable material and air. The flammable cloud then burns as a "diffusion" flame in which the rate of energy release is determined largely by mixing, predominantly at that stage a buoyancy effect. Because combustion is always, by definition, rich of stoichiometric, considerable quantities of carbon are formed–when hydrocarbons are involved–and thermal radiation levels are high. By the same token combustion is always incomplete.

UVCEs, on the other hand, would appear to be phenomena in which mixing prior to combustion occurs and indeed is necessary. Kinetic energy in the form of blast waves affects areas far beyond the boundaries of thermal damage. Possibly an appreciable fraction of the total combustion energy of that part of a flammable release, which is in explosible admixture with air, can manifest itself as kinetic-blast energy, apparently as much as 60% (Table 1).

Both fire-balls and UVCEs are included here, because the circumstances which give rise to the former are valuable in defining, by comparison, what are probably the crucial factors in the formation of the latter. Whereas blast effects have, in general, led to the identification of most UVCEs, some of which have involved surprisingly small quantities of flammables, fire-balls of a similar scale are commonplace in most fires involving highly flammable materials and, except for the largest, have gone unrecorded.

The latest publishable information on these events is presented in Table 1. This information has been derived from a number of sources, of which the latest and most comprehensive are those of Strehlow[43] and Davenport[44]. Some of the events listed are obliquely referred to in the open literature[17], but have been expanded by

Table 1 – *Summary of data for some principal UVCEs*

No	Date	Place	Flammable	Source	Quantity	Blast	TNT Equivalent	Fatalities	Ref No
1	23 Aug 1921	Hull, England	Hydrogen	Dirigible ZR-2	6900 kg	Yes	450 kg Yield: 0.25%	1	44 47 48
	Windows broken within 3.2 km area. Blast heard 8.04 km away. Tremors felt 80.4 km								
2	17 Dec 1932	Detroit, Mich, USA	LPG	Storage	–	No	–	0	43 49
	50 mm temporary hose connected to 83 m^3 above-ground storage vessel broken								
3	2 June 1934	Huntington Beach, Calif, USA	Butane	Road tanker	8600 kg	No	–	0	43 50
	Broken 75 mm connection on 19 m^3 LPG container escape lasted 5 minutes before ignition								
4	22 Oct 1936	Crowley, La, USA	Butane	Road tanker	–	No	–	4	43 51
	Snapped 50 mm pipe connection. Vapours travelled 50 m and then flashed								
5	2 Jan 1939	Newark, NJ, USA	Butane	–	–	Yes	?	0	43 52
	Vapours accumulated in open courtyard between three buildings. Doors and windows blown in								
6	11 June 1941	Los Angeles, Calif, USA	Butane	Road tanker	–	No	–	0	53
	8.7 m^3 LPG trailer ran away and was holed upon collision. Vapours extended over a wide area								
7	18 Jan 1943	Los Angeles, Calif, USA	Butane	Road tanker	–	No	–	5	43 44 55
	Bottom connection on 16.2 m^3 tank failed, allowing liquid butane to escape, creating 0.4 km diameter cloud								
8	10 Aug 1943	Home Gas Corp, Palmer, Mass, USA	Propane	LPG storage	4300 kg	No	–	0	43 54
	Broken 50 mm pipeline escape lasted for 1½ minutes before line isolated and then contents of line vented. Propane pressure 1.72 MPa								
9	15 Nov 1944	Cleveland, Ohio, USA	Methane	LNG storage	182 000 kg	No	–	213	43 56
	Cryogenic tank containing 4500 m^3 LNG developed a leak. Methane spread widely before ignition								
10	21 Nov 1944	Denison, Tex, USA	Butane	LPG storage	–	No	–	10	43 57
	An escape lasting 2 minutes occurred on a 3.8 m^3 tank before ignition. Everyone within 90 m burned								
11	25 April 1945	Los Angeles, Calif, USA	Butane	LPG storage	–	Yes	?	0	58
	Escaping vapours ignited 37 m from their source: "Explosion shattered buildings"								
12	1945	NJ, USA	Crude oil	Refinery storage tanks	–	No	–	0	4[illegible]
	Failure of cast iron valve on storage tank created cloud which was ignited more than 60 m away								
13	13 Oct 1948	Sacramento, Calif, USA	Butane	Road tanker	–	No	–	2	5[illegible]
	Butane tanker struck by train								

'able 1 – *Summary of data for some principal UVCEs (continued)*

o	Date	Place	Flammable	Source	Quantity	Blast	TNT Equivalent	Fatalities	Ref No
4	28 July 1948	BASF, Ludwigshafen, W Germany	Dimethyether	Rail tank car	30 000 kg	Yes	–	209	44 59a
	Rail car "bubble-full" overheated in sun. No pressure relief valve fitted. Car ruptured explosively. Dispersed contents formed vapour cloud before ignition								
5	23 June 1949	California Ref Co, Perth, NJ, USA	Hydrocarbons	Storage tank	160 000 kg	No	–	4	60
	317 m^3 tank rocketted after failure of bottom seam								
	27 Oct 1949	Winthrop, Mo, USA	LPG	Rail tank car	–	No	–	1	43 61
	Accident whilst unloading								
	23 Aug 1950	Wray, Colo, USA	Propane	Road tanker	–	No	–	2	43 62
	10.5 m^3 capacity tanker overturned, snapping off pipe connection. Vapour cloud developed, which subsequently ignited								
	7 Oct 1950	Woodbury, Ga, USA	LPG	Storage tank	–	No	–	0	43 63
	Pipe break on 3.8 m^3 capacity storage tank								
	6 March 1951	Koppers Co Inc, Kobuta, Pa, USA	Butadiene	Process	–	No	–	0	43 64
	Leak occurred in 12.5 mm bleeder valve, which ignited 56 m away								
	2 Jan 1952	–	Propane	Rail tank car	–	No	–	0	65
	Accident led to 600 mm hole in tank car. "Explosion" occurred but no blast reported								
	24 June 1952	Kansas City Kans, USA	LPG	Road tanker	–	Yes	?	2	66
	Vapour discharged through relief valve. Explosion heard 3.2 km away								
	21 July 1952	Bakersfield, Calif, USA	Butane	Storage	–	No	–	0	43 67
	2 ha cloud formed after earthquake								
	Jan 1954	Tenn, USA	Isopropyl alcohol	Process	–	No	–	0	44
	Vapour cloud formed from unknown leak source								
	18 Oct 1954	Portland, Oreg, USA	LPG	Rail tank car	–	Yes	?	0	44 68
	60 m vapour spread after safety relief valve stem broke on tank car. Storage tanks 18 m away slightly distorted. 'Flames transformed to explosion after travelling a few feet'								
	1954	W Va, USA	Acrolein	Rail tank car	18 200 kg	Yes	?	0	44
	Tank car burst due to runaway polymerisation of contaminated acrolein								
	22 July 1955	Wilmington, Calif, USA	Butane	Process	–	Yes	?	0	44 69
	Rupture of 150 mm pipe containing butane at 3.5 MPa gave explosion and fire								

Table 1 – *Summary of data for some principal UVCEs (continued)*

No	Date	Place	Flammable	Source	Quantity	Blast	TNT Equivalent	Fatalities	Re Nc
27	1956	New York, USA	Ethylene	Polyethylene plant	1340 kg	Yes	?	0	44
	Release from polyethylene reactor at 136 MPa. Aerial explosion created 3.4 kPa overpressures								
28	8 Jan 1957	Montreal East, Canada	Butane	Storage	–	Yes	?	1	70
	Vapour cloud formed during filling operation – ignited 180 m from leak								
29	1957	Quebec, Canada	Butane	Storage	–	No	–	1	44
	Storage sphere failed, creating 366 m diameter cloud								
30	24 Oct 1957	Sacramento, Calif, USA	LPG	Road tanker	–	Yes	?	1	71
	Coupling snapped. Vapour ignited "and exploded" very soon after emission started								
31	1958	Mich, USA	Butane	Refinery	–	No	–	1	44
	Overpressure of 80 m^3 tank caused failure. Cloud ignited 107 m from source								
32	1958	Calif, USA	Oil froth	Refinery	–	No	–	2	4
	Water in 7900 m^3 tank caused frothing. Amount of vapour formation not known								
33	15 April 1958	Ardmore, Okla, USA	Propane	Refinery	–	Yes	?	0	4 7
	Coupling on tank car failed. Workmen blown into ditch								
34	3 July 1958	Boron, Calif, USA	LPG	Storage	545 kg	No	–	1	7
	Escape from 3.8 m^3 storage tank through broken connection								
35	30 July 1958	Augusta, Ga, USA	LPG	Road tanker	–	Yes	?	1	7
	Escape during filling. Vapours spread widely. Although fire was confined 5 houses were destroyed and 85 othe damaged								
36	28 June 1959	Meldrin, Ga, USA	LPG	Rail tank car	–	Yes	?	23	4 4 7
	Train derailed on trestle bridge in country close to picnic area. Tank car punctured and gases spread over a wid area before ignition. Explosion heard 3.2 km away								
37	1961	Tex, USA	Cyclohexane	Process	3860 kg	Yes	~25 kg Yield: 0.06%	1	4
	Valve failed at bottom of reactor. Control room crushed. Estimated overpressure 13.7 kPa								
38	17 April 1962	Olin Mathieson, Doe Run, Ky, USA	Ethylene oxide	Storage	22 000 kg	Yes	18 200 kg Yield: 18%	1	[illegible]
	Tank containing ethylene oxide became contaminated with ammonia. Tank ruptured, dispersing ethylene oxid into air. Cloud immediately ignited								
39	25 July 1962	Berlin, NY, USA	LPG	Road tanker	12 500 kg	Yes	?	10	[illegible]
	Cloud 122-183 m in diameter, 24 m high, exploded. Low overpressures created								

able 1 – *Summary of data for some principal UVCEs (continued)*

Date	Place	Flammable	Source	Quantity	Blast	TNT Equivalent	Fatalities	Ref No
4 Aug 1962	Middle East	Propane	Storage	1140 kg	Yes	?	1	43 78
Vapours from refrigerated propane storage escaped from tank which ruptured due to overpressure. Personnel 90 m away blown over by force of explosion when vapours ignited								
1963	Dow Chemical Plaquemine, La, USA	Methane or ethylene	Process	–	Yes	1000 kg	0	22
Sight glass or metering station failure. Explosion of cloud about 30 seconds after failure								
9 Jan 1964	NRDS, Jackass Flats, Nevada, USA	Hydrogen	Experiment	90 kg	Yes	27 kg Yield: 1%	0	79
Accidental ignition of hydrogen during venting experiment								
1964	Tex, USA	Ethylene	Process	90-230 kg	Yes	120-300 kg Yield: 13%	2	44
Failure of compression fitting on 7-17 MPa ethylene line in pipe trench. Explosion in courtyard between buildings giving partial confinement								
13 July 1965	Cities Service Oil Co, Lake Charles, La, USA	Methane and ethylene	Process	–	Yes	?	0	80
Vented cold liquid into carbon steel flare header. Extensive explosion damage								
1965	Ethyl Corp, Baton Rouge, La, USA	Ethyl chloride	Process	17 300 kg	Yes	150-200 kg Yield: 0.25%	0	44
Rupture of recycle line on reactor								
24 Oct 1965	Escambia Chem Corp	Hydrogen and carbon monoxide	Process	70-140 kg	Yes	12 kg Yield: 1%	0	81
Valve failure led to discharge at 35 MPa through 25 mm diameter opening for 10-20 seconds before ignition. Damage up to 240 m away								
1965	Tex, USA	Butane	Process	–	No	–	0	44
Removal of wrong bind flange led to spill of unknown quantity. Ignited by heater 18 m from spill point								
4 Jan 1966	Feyzin, France	Propane	Storage	432 000 kg	No	–	17	82
Rupture of 1200 m^3 capacity storage sphere whilst under fire attack								
16 Jan 1966	Caltex, Raunheim, W Germany	Methane	Process	500 kg	Yes	1000-2000 kg Yield: 18-36%	1	17 18 44
Liquid methane–vented at high level–subsequently formed a low lying dispersion. Blast damage to plant not particularly severe. Extensive glass breakage at 400 m–minor breakage at 1200 m								
Jan 1966	W Germany	Ethylene	Pipeline	–	Yes	2000-3000 kg	0	17
Gas reached 60 m before ignition. Ethylene plant control room situated 25-30 m from plant rendered inoperative –laminated safety glass windows blown in								
1966	La, USA	Butadiene	Reactor	385 kg	Yes	?	3	44
Valve on reactor bottom opened accidentally. Ignition source 15 m from point of spill. Overpressures of 3.4–6.9 kPa were estimated								

2: Although this would appear to be one of the more clear-cut cases, recent work has indicated that this might so. This case is discussed further in the body of the text

Table 1 – *Summary of data for some principal UVCEs (continued)*

No	Date	Place	Flammable	Source	Quantity	Blast	TNT Equivalent	Fatalities	Re Nc
52	1966	W Germany	Light HC's	Refinery	–	No	–	45	44
	Frozen open water draw-off valve on bottom of storage sphere resulted in cloud which was ignited by vehicle 100 m away. Sphere subsequently ruptured								
53	23 May 1966	Gulf Ref Co, Pa, USA	Cumene, benzene and propane	Refinery	–	No	–	0	83
	Materials discharged at 3.4 MPa from rupture of 150 mm pipe. Cloud was wind affected								
54	8 Aug 1967	Cities Service Oil Co, Lake Charles, La, USA	Isobutylene	Process	9100 kg	Yes	10 000–12 000 kg Yield: 11-13%	7	44 84
	Failure of 250 mm gate valve bonnet. Explosive force estimated from glass breakage								
55	20 Jan 1968	Shell Netherlands, Pernis Holland	Light HC's	Storage	50-100 tonne	Yes	20 tonne Yield: 4-8%	2	8 8
	Breaking of water-oil emulsion in slop-oil tank caused huge cloud								
56	1968	Tex, USA	Light HC's and acid	Process	–	No	–	1	4
	Cloud formed when 250 mm spool piece split longitudinally. Ignited at furnace 55 m away								
57	1968	La, USA	Vinyl chloride	Process	–	No	–	0	4
	Failure of pump created cloud which was ignited at furnace 30 m away. Cloud less than 1 m high in most areas								
58	5 Dec 1968	Yutan, Neb, USA	LPG	Pipeline	–	No	–	5	8
	Gas escape isolated before ignition. Ignited–when repair crew arrived 6 hours later–at a point about 30 m away from the break								
59	28 Dec 1969	Fawley, England	Naphtha and hydrogen	Process	23 000 kg	Yes	?	0	4
	Failure of 400 mm pipe at 1.7 MPa. Ignition within 10 seconds								
60	11 Sept 1969	Black Bayou Jct, Miss, USA	Vinyl chloride	Rail tank car	–	No	–	0	8
	Escape after derailment. Ignition 6 hours later								
61	14 May 1969	ICI, Wilton, England	Cyclohexane	Process	–	No	–	2	4
	A low cloud formed from 4 tonne release at 689 kPa from an improperly blinded pipe. Ignition was by a diesel engine 30 m from spill point								
62	1970	NJ, USA	Heavy HC's and hydrogen	Process	114 000 kg	Yes	50 000 kg Yield: 4%	0	4
	Failure of reactor operating at 17.2 MPa due to localized overheating. Overpressures were highly directional. Peripheral damage indicated the explosion yield								
63	2 June 1970	Crescent City, Ill, USA	–	Rail tank cars	–	No(?)	–	0	4 8
	Train derailment caused ruptures								
64	23 Oct 1970	Hull, England	Propane	Road tanker	200 kg	No	–	2	5

'able 1 – *Summary of data for some principal UVCEs (continued)*

o	Date	Place	Flammable	Source	Quantity	Blast	TNT Equivalent	Fatalities	Ref No
5	12 Nov 1970	Hudon, Ohio, USA	LPG	Road tanker	–	No	–	6	90
	Vehicle driven away whilst still connected by hose to a fixed installation. 50 mm hole gave rise to liquid discharge								
6	9 Dec 1970	Franklin Co, Port Hudson, Mo, USA	Propane	Pipeline	70 000 kg	Yes	50 000 kg Yield: 7.5%	0	91
	Spill from a pipeline rupture produced cloud 460 m long, and 3–6 m high								
	26 Feb 1971	Longview, Tex, USA	Ethylene	Process	–	Yes	?	3	92
	Discharge from 276 MPa polymerisation reactors. Explosion felt 9.6 km away								
	19 Jan 1971	La, USA	Ethylene	Road tanker	3640 kg	Yes	500 kg Yield: 1%	0	44
	Failure of frozen relief valve at about 2.8 MPa. Widespread damage to light construction components								
	1971	Tex, USA	Ethylene	Process	450 kg	Yes	?	3	44
	Three explosions occurred. Some confinement from building walls around an alleyway								
	Sept 1971	Tex, USA	Butadiene	Process	12 300 kg	Yes	?	1	44
	Failure of pump. 19 km/h wind caused cloud to drift 180 m before ignition from unknown source 8–10 minutes after release								
	22 Jan 1972	East St Louis Ill, USA	Propylene (94%)	Rail tank cars	54 000 kg	Yes	1000-2500 kg Yield: 0.2-0.5%	0	44 86
	Rail cars collided in humping operation in switchyard, creating an elongated vapour cloud covering about 2 ha. Two explosion centres were identified; there "may have been a detonation"								
	1972	Brazil	Butane	Storage	–	Yes	?	37	44
	Frozen open water draw-off valve on bottom of storage sphere								
	14 May 1972	Exxon, Hearne, Tex, USA	Crude oil	Pipeline	640 000 kg	No	–	1	44
	Split in 150 mm pipeline caused discharge which lasted 4½ hours before ignition. Scorched area 550 m long. Material at 3.6 MPa when released								
	Sept 1972	France	LNG	Experiment	–	No	–	0	86
	1972	Montana, USA	Butane	Process	–	No	–	1	44
	Leak from partly open 150 mm flange was ignited at furnace 30 m away								
	19 March 1972	Lynchburg, Va, USA	Propane	Road tanker	1500 kg	No	–	1	86
	Delay to ignition: 3–4 minutes								
	1973	Japan	Ethylene	Process	–	Yes	?	1	44
	Excessive pressure and temperature from process upset caused valve flange to fail								
	28 Oct 1973	Shinctsu, Japan	Vinyl chloride	Process	4200 kg	Yes	?	1	44 59a
	Broken valve yoke								

Table 1 – *Summary of data for some principal UVCEs (continued)*

No	Date	Place	Flammable	Source	Quantity	Blast	TNT Equivalent	Fatalities	Ref No
79	23 Feb 1973	Austin, Tex, USA	LNG	Pipeline	–	No	–	8	86
	Delay to ignition: 15 minutes								
80	Jan 1974	Florida, USA	Propane	Road tanker	16 400 kg	Yes	?	0	44
	Hose from tanker failed, causing 12 m high vapour cloud. Explosion destroyed two warehouses, crushed cars and broke windows in four block area								
81	1 June 1974	Flixborough, England	Cyclohexane	Process	36 tonne	Yes	18 tonne Yield: 5%	28	20 21 93
	Poorly installed 500 cm temporary pipe failed. Ignition in 25–35 seconds created large overpressures								
82	29 June 1974	Climax, Tex, USA	Vinyl chloride		75 000 kg	Yes		0	56
83	19 July 1974	Decatur, Ill, USA	Propane	Rail tank car	63 tonne	Yes	20-40 tonne Yield: 32-65%	7	44 86
	Accident created 56 cm × 65 cm hole in end of rail car, releasing contents. Ignition of the 800 m × 1200 m cloud was 8–10 minutes after the collision. An unusually high yield was indicated, based on damage								
84	25 Aug 1974	Petal, Missouri, USA	Butane	Storage	–	Yes	?	0	44 94
	Overfilling of salt dome storage well created a cloud 2 km in diameter. Two explosions occurred. The second was 245–300 m above grade and was the most severe. Glass breakage to 11.3 km. Houses damaged to 275 m								
85	5 Sept 1974	Solvay et Cie, Martorellas Plant, Barcelona, Spain	Dichlorethylene and vinyl chloride	Process	22 tonne	No	–	1	95
	Pipe rupture led to discharge of a mixture of DCE and VC in the ratio 7 : 3 at a pressure of 1.1 MPa and a temperature of 150°C. Ignition occurred 75 seconds after commencement of discharge								
86	1974	England	Ethylene	Process	910-2700 kg	Yes	7 kg +	0	44
	Leak from high pressure piping caused spill lasting about 30 minutes before ignition. Windows broken in 180 m radius. Roofs and ceiling in four buildings damaged								
87	13 Sept 1974	Griffith, Ind, USA	Propane	Storage	> 100 tonne	No	–	0	86
	45 cm vertical pipe at 10 atmospheres connected to 36 000 m^3 cavern discharged for 7 hours before ignition occurred with a few "pops"								
88	21 Sept 1974	Houston, Tex, USA	Butadiene	Rail tank car	Not more than 80 tonne	Yes	20-57 tonne Yield: 3-7%	1	44
	Humping accident in railyard punctured tank car. Amount of spill in 2-3 minutes before ignition not known. Ignition by locomotive 180 m away. Estimate 14-21 kPa overpressure at 300 m from point of rupture								
89	1974	Tex, USA	Pentanes	Process	7600 kg	Yes	1300 kg Yield: 2%	2	44
	Piping failure. Explosion knocked out deluge systems and fire pumps. Maximum overpressure was 24 kPa at 37 m								
90	18 July 1974	Dow Chemical Co, Plaquemine, La, USA	Propylene	Process	–	No	–	0	96
	Failure of expansion joint followed by ignition at furnaces. One deluge system between spill point and ignition source tripped prior to ignition								

Table 1 – *Summary of data for some principal UVCEs (continued)*

No	Date	Place	Flammable	Source	Quantity	Blast	TNT Equivalent	Fatalities	Ref No
1	10 Feb 1975	Union Carbide, Antwerp, Belgium	Ethylene	Process	2500 kg	Yes	?	6	44 96a
	Failure of a branch connection on a liquid ethylene header in a high-pressure pump room. Discharge through 25 mm opening for about 1 minute before ignition. No source of ignition was identified. Explosion appeared to be highly directional: 34–48 kPa at 60 m in one direction, 3.4–6.9 kPa at 60 m in opposite direction								
2	7 Nov 1975	DSM, Beek, Holland	Propylene	Process	5450 kg	Yes	2200 kg Yield: 4%	14	97
	Failure of level controller caused cold liquid to pass into carbon steel flare header which cracked. Ignition of cloud at furnace 46 m away								
[illegible]	1975	W Germany	Naphtha plus hydrogen	Process	–	Yes	?	0	44
[illegible]	1975	Calif, USA	Hydrogen	Storage	300 kg	Yes	10-20 kg Yield: 0.1–0.2%	0	44
	1975 or 1976	Czechoslovakia	Light HC's	?	–	Yes	?	14	44
	No details available								
	1976	Tex, USA	Ethylene	Pipeline	–	Yes	?	1	44
	Line failure at 13.8 MPa								
	Feb 1976	Tex, USA	LPG	Pipeline	–	No	–	1	44
	Crosscountry pipeline failure								
	1976	Puerto Rico	Pentanes	Storage	–	No	–	1	44
	Tank overpressure led to release of vapour or mist which was 4.6 m high. Ignition was at a furnace 137 m from the tank. Scorched area was about 74 m^2. Cloud was consumed in 10–30 seconds								
	21 Sept 1976	NJ, USA	Propylene	?	18 200 kg	No	–	2	56
[illegible]	20 Feb 1977	Dallas, Tex, USA	Isobutane	Rail tank car	68 200 kg	Yes	1600 kg Yield: 0.25%	0	98
	Tank car punctured during derailment. Delay to ignition about 2½ minutes. The "open-air explosion was extremely violent and the shock was felt for miles around". Locomotives would appear to have been exposed to about 3.5 kPa (photograph) at a distance of about 300 m from punctured tank car								

private communication. For legal or commercial reasons some recent significant events have had to be omitted. So far as can be judged, where sufficient of the facts are known to me, the omitted incidents do not indicate behaviour disparate from what has come to be expected from UVCEs, although some exhibit magnitudes in excess of any tabulated herein.

Table 1 reveals that there have been up to February 1977 at least 100 major flammable gas or vapour releases, which have ignited. 56 of these have produced blast effects of varying magnitudes. The recorded incidents span 55 years, but more than half have occurred in the last 12 years (to September 1977) and nearly 60% of these were associated with blast.

LPG and its principal constituents, propane and the butanes, figure largely in the table (43 incidents, 18 giving rise to blast), particularly in the earlier years. Butadiene, isobutylene and propylene as pure single components are also represented in nine incidents, of which six gave rise to blast—as are "pure" methane (LNG) and "pure" hydrogen (four and three incidents, one and three blasts, respectively). Others represented as "pure" single-component vapours are:

dimethyether	1 incident,	1 blast;
acrolein	1 incident,	1 blast;
isopropyl alcohol	1 incident,	0 blast;
cyclohexane	3 incidents,	2 blasts;
ethylene oxide	1 incident,	1 blast;
ethyl chloride	1 incident,	1 blast;
vinyl chloride	4 incidents,	2 blasts.

The only "pure" material appearing in Table 1 which seems to have been associated with blast exclusively, however, is ethylene—ten incidents, each of which resulted in blast. One unknown material gave no blast (No 63). Many of the escapes were mixtures, the combustion characteristics of which might have been determined by the more reactive constituent(s). LPG has already been mentioned; the others are:

crude oil	2 incidents,	no blasts;
"light" hydrocarbons (including pentanes)	8 incidents,	4 blasts;
"oil froth"	1 incident,	no blast;
methane *or* ethylene	1 incident,	1 blast;
methane *and* ethylene	1 incident,	1 blast;
dichloroethylene and vinyl chloride	1 incident,	no blast;
hydrogen and carbon monoxide	1 incident,	1 blast;
hydrogen and naphtha	2 incidents,	2 blasts;
hydrogen and "heavy" hydrocarbons	1 incident,	1 blast.

In those cases where the initial flammable material also contained hydrogen—Nos 59, 62 and 93—or ethylene—No 44 and perhaps No 41—blast resulted. Of the 27 incidents which did involve (or might have involved) unsaturated flammable materials, 21 gave rise to blast effects.

There would appear to be a distinct correlation between the occurrence of blast and the involvement of unsaturated materials, although again caution has to be exercised in the interpretation, as most of the ethylene bangs occurred as the result of accidental or deliberate dumping from very-high-pressure equipment.

The incidents have the following common factors: they involved large releases to the atmosphere of flammable materials at moderate to high rates; the releases were all into spaces where (unless otherwise specified) the only large, coherent, solid surface near the vapour cloud was the ground. In most cases the releases were not close to sources of ignition and, again in most cases, ignition delays, although variable, were considerable, extending from several seconds to hours.

There were difficulties in preparing Table 1. Davenport[44] has already highlighted these and I have accepted his data and deductions, supplemented by a small number from elsewhere, as probably the best that currently exist. Clearly much relevant information was contained in the 40 private communications which supplemented the 56 published sources he quoted. Indeed, 14 of those 56 proved inaccessible to me. Eighteen of the accessible sources are journals primarily aimed at Fire Brigade personnel and largely prepared by Fire Brigade staff. Only 15 references cited truly technical or authoritative sources.

Both the Fire Brigade and the technical sources devote greater attention to causes than effects and only in very few instances has there been any systematic study of the damage; opinions are rarely supported by data, figures by calculation, or descriptions by photographs.

Wherever possible I have tried to identify the precise date of the incident referred to and the precise location of the happening. Lack of time and inaccessible original sources have prevented me from accomplishing this in many cases. The column headings in Table 1 are in the main self-explanatory. "Quantity" figures are those for either the total quantity of *vapour* involved in any given spill before ignition, or the total quantity spilt (both vapour and liquid) where sufficient time elapsed, before (or during) ignition, for substantial evaporation to have occurred. "Quantity" (where quoted) is considered to be reliable only for total mass emitted, but possibly to the extent of no better than a factor of two in most cases. The proportion of that quantity as vapour in explosible admixture with air is clearly susceptible to much greater uncertainty.

TNT equivalents correspond to damage having a close resemblance to that produced by high explosive at the same distance from the cloud centre. For the assessment to be considered in any way realistic damage patterns must match and distances should be accurately assessed. Naturally, accurate assessment of distances from the "centre" of a large ill-defined volume of flame such as typically constitutes an UVCE presents insuperable difficulties. Often the type of damage found can only be correlated approximately with damage occasioned by explosives. The differences found correspond to substantial differences in forms and durations of pressure waves; see the Appendix. The only "reliable" correlations with explosives are those in the far-field where the pressure wave has generated a steep (shock) front. Unfortunately (from a utilitarian point of view) the far-field is also, in most cases, associated with degeneration of the shock wave to such a low amplitude that it no longer causes diagnostic damage.

The values of "yield" (or efficiency) quoted in the Table are those based upon the equivalent quantity of TNT—which would have produced similar damage patterns—and the quantity of the flammable material involved having regard to the energy release of the materials upon combustion or detonation. Yield is clearly widely variable, depending so much upon what is assumed or assessed to be the quantity of flammable in explosible admixture with air and the extent of reaction. Ideally, TNT equivalents should be based upon "far-field" effects but often are not, which leads to uncertainty and incomparability. Yield is assessible with accuracy, to say better than 10% in terms of the volume or mass of flammable involved, in well-defined circumstances only, *ie*, No 42 (but see Section V). In most of the 22 cases, where yield has been calculated, even the total quantity escaping before ignition has proved incapable of assessment within a factor of two and the best that can be said about the quantity in admixture with air and within the limits of flammability is that it was less than the total involved. How much less is in my view susceptible to an error amounting to possibly an order of magnitude. These yields, therefore, are to be seen as representing less than maximum "efficiencies" and are not to be regarded necessarily as realistic indications of average or typical behaviour. Methane, for example, could theoretically on a purely thermal basis produce damage equivalent to the detonation of 10.94 times its own weight of TNT (calculated by dividing the net heat of combustion of methane–11 954 k cal/kg–by the blast energy output of TNT–1093 k cal/kg). This is the 100% efficiency (yield) criterion. It is difficult to envisage this being exceeded. Indeed, it is probable that less than the whole combustion heat would be available to produce kinetic energy (as is precisely the case with TNT[45,46]), for the following reasons (not all of which would necessarily apply): slowness of certain reaction stages, indadequate mixing or flammable-rich concentrations and dissociation. It is further likely that the former and the latter would only assume significance when flame speeds approach detonation velocities. It will be appreciated that this comment begs the question, since whether or not UVCEs involve detonation is still a live issue. If UVCEs arise as a result of ordinary flame processes then the yield, or efficiency, of the process would probably be *greater* than that corresponding to detonation by a factor of about 1.25[23]. In other words, a detonative UVCE would not be expected to exceed 80% efficiency.

Table 2 – *Combustion properties of the flammables in some of the UVCEs documented in Table 1, where blast occurred*

No	Flammable	Quantity	Yield	Combustion Heat Release	Combustion Limits		Maximum Flame Speed	Value of Combined Function: Columns 3, 5, 6 and 7 Multiplied	Source Pressure	State of Flammable on Emission			State of Cloud at Time of Ignition	Principal Orientation of Discharge
		kg	%	kcal/kg	UL	LL	m s^{-1}	MW-m-%	MPa	G	SL	CL	Phase	
1	Hydrogen	6900	0.25	28 670	75	4	2.65	1.6×10^8	Low	x			Single	3 dimensional
37	Cyclohexane	3860	0.06	10 480	9.1	1.3	0.43	5.7×10^5	Medium (?)		x		Two	Horizontal
38	Ethylene oxide	22 000	18	6390	100	3	1.00	5.7×10^7	Medium		x		Two	3 dimensional
42	Hydrogen	90	1	28 670	75	4	2.65	2.0×10^6	23.5	x			Single	Vertical
43	Ethylene	90/230	13	11 270	32	3.1	0.64	$7.8 \times 10^4/2.0 \times 10^5$	6.9/17.3	x or	x		Single	?
45	Ethyl chloride	17 300	0.25	4580	12	3.6	?	–	Medium (?)		x(?)		Two	?
46	Hydrogen/ carbon monoxide	70/140	1	28 670/2400	75	4	2.65/0.40	–	34.5	x			Single	?
49	Methane	500	18/36	11 950	15	5	0.37	9.2×10^4	Low			x	Two	Vertical
54	Isobutylene	9100	11/13	10 800	9.7	1.6	0.48	1.6×10^6	Low		x		Single (?)	Horizontal
55	Light H/C s	50 000/100 000	4/8	10 400	~ 9	~1.4	~0.40	$6.6 \times 10^6/1.3 \times 10^7$	Low	x			Two	Vertical
62	Heavy H/C s and H_2	114 000	4	10 400	~47	~0.6	~0.40	9.2×10^7	17.3		x		Two	?
66	Propane	57 000	9	10 900	9.5	2.2	0.43	8.2×10^6	6.5		x		Two	Horizontal
68	Ethylene	3640	1	11 270	32	3.1	0.64	3.2×10^6	2.8	x or	x		Two	?
71	Propylene	118 000	0.2/0.5	10 940	10.3	2.4	0.48	2.1×10^7	Low		x		Two	Horizontal
81	Cyclohexane	36 000	3/6	10 480	8	1.3	0.43	4.5×10^6	0.86		x		Two	Horizontal
83	Propane	63 000	32/65	10 900	9.5	2.2	0.43	9.0×10^6	1.0		x		Two (?)	Horizontal
86	Ethylene	900/2700	0.02+	11 270	32	3.1	0.64	$7.8 \times 10^5/2.4 \times 10^6$	High	x or	x		Two (?)	?
88	Butadiene	<80 000	3/7	10 820	13.3	2	0.48	2.0×10^7	0.2		x		Two	Horizontal
89	Pentanes	7600	2	10 820	9	1.4	0.42	1.1×10^6	Low (?)		x(?)		?	?
92	Propylene	5450	4	10 940	10.3	2.4	0.48	9.5×10^5	Low (?)			x	Two	Vertical
94	Hydrogen	300	0.1/0.2	28 670	75	4	2.65	6.8×10^6	?	x			Single	Vertical
100	Isobutane	68 200	0.25	10 840	8.4	1.8	0.42	8.6×10^6	Low		x		Single	Horizontal

"Low" means 0 to 690 kPa
"Medium" means 690 to 6900 kPa
"High" means 6.9 MPa or more

"UL": upper explosive limit, volume percent
"LL": lower explosive limit, volume percent

"G": gas
"SL": superheated liquid with reference to atmospheric condition
"CL": cryogenic liquid

Table 2 is a distillation of what seem to be the salient factors from Table 1 in cases where blast occurred. This Table attempts to correlate factors such as energy release, flame speed, force of discharge, state of combustible on discharge and quantity discharged prior to ignition *etc* with yield. Figures 1 and 2 show two possible presentations of the data—those of yield *vs* quantity of flammable, and yield *vs* the total product of quantity of flammable, heat of combustion, flammable range and burning velocity.

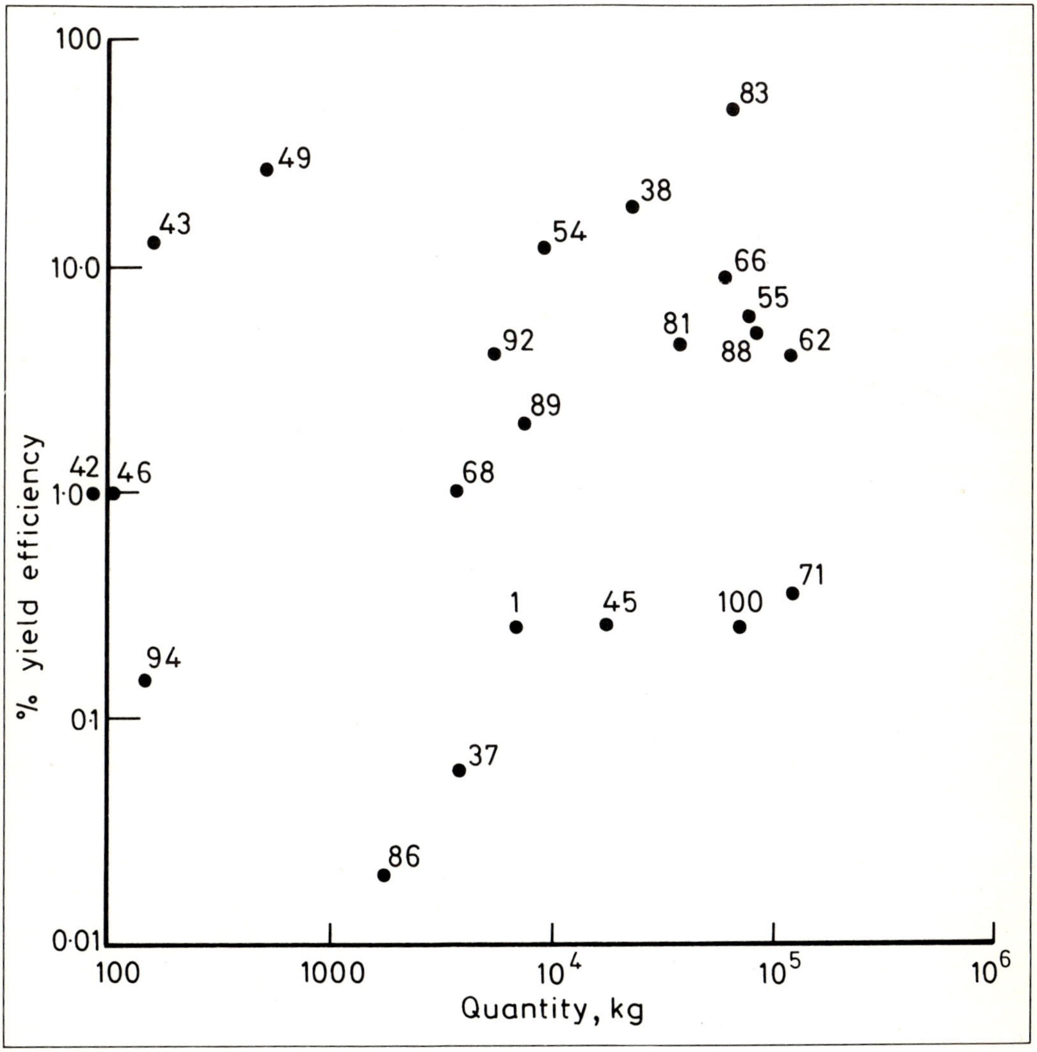

Figure 1 – *A 4-cycle by 4-cycle logarithmic plot of the assessed efficiency of 22 unconfined vapour cloud explosions, expressed as a percentage of the equivalent yield as TNT against the quantity of flammable material involved in kilograms. The figures adjacent to the points refer to case numbers assigned to the incidents in the Tables and the text*

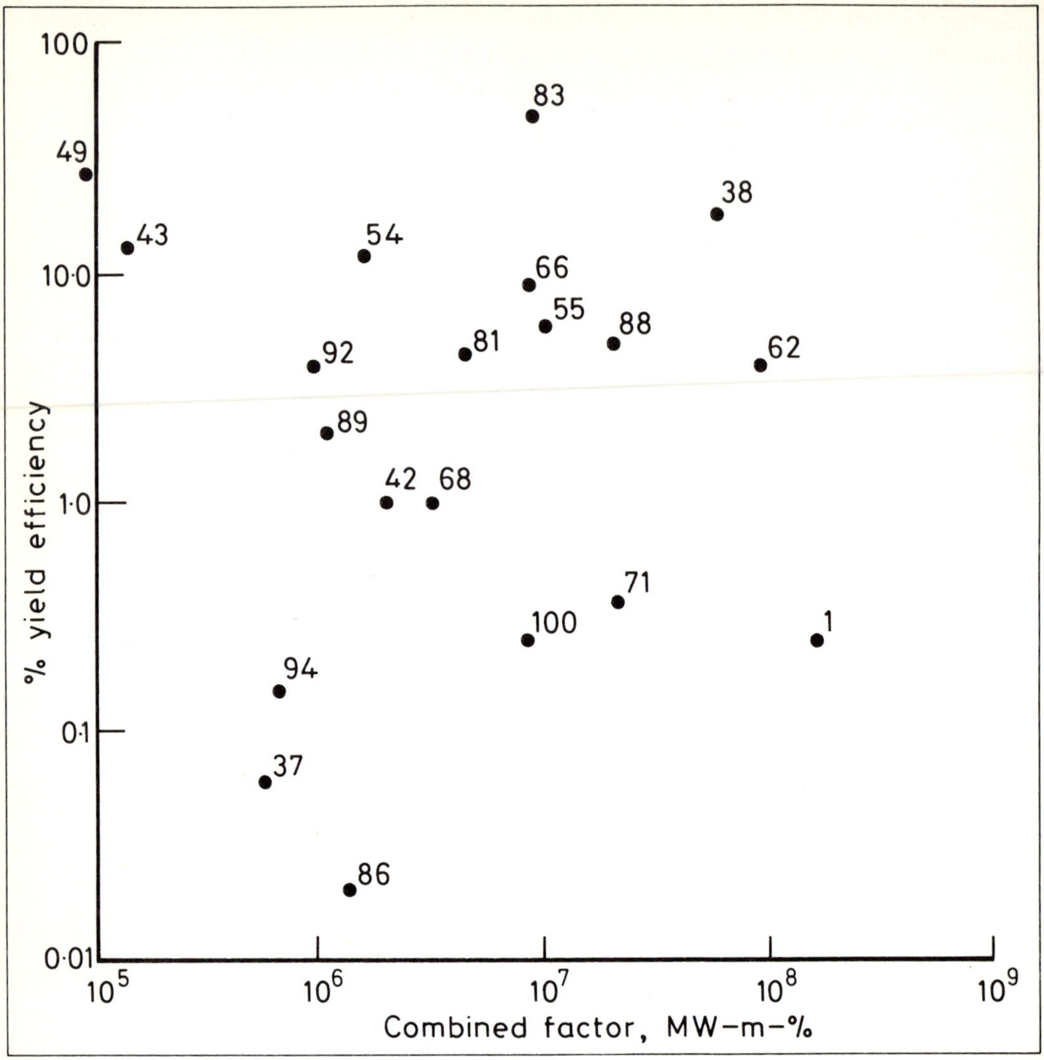

Figure 2 – *A 4-cycle by 4-cycle logarithmic plot of the assessed efficiency of 20 unconfined vapour cloud explosions, expressed as a percentage of the equivalent yield as TNT against a composite factor formed from the product of quantity of flammable, heat of combustion, range of flammability and fundamental burning velocity, the composite factor being expressed in MW-m-%*

To encompass all the cases in Table 2 a 4-cycle by 4-cycle log-log display is necessary, from which it is clear that no straightforward correlations on the chosen bases are immediately apparent. Linear correlations would be unsuitable as they would intersect 100% yield–the *a priori* limit of blast efficiency–at finite quantities of flammable. All monotonic correlations would appear to exhibit scatter beyond the limits which would render any such correlation acceptable for predictive use–the real purpose of this research study.

The data on fatalities is used later in an endeavour to identify the actual risk of loss of life in any particular case. This, however, is not a study of the probability of fatality, which is dependent upon additional factors outside the scope of this survey—reliability, quality of personnel, *etc.* Correlation with non-fatal injury has not been attempted, as this depends upon the definition of "injury" and is in any event a figure even more variable than fatality. In all the incidents cited the total of fatalities was at least 725, of which 385 died in those incidents associated with pressure effects. In the last 12 years 226 have died, some 58% of whom (132) were involved in UVCEs.

Both the frequency of all incidents and the numbers of fatalities reveal an increasing trend which may be rationally associated with larger numbers of plants and with increased inventories—the "big single stream syndrome"—in recent times. It would appear, however, that the occurrence of UVCEs has not significantly increased as a proportion of the total. Neither does it unequivocally appear that fatalities are significantly greater in UVCEs. This is due to the large polarising influences of two events: the Cleveland National gas disaster (No 9) and the Ludwigshafen explosion (No 14).

Many of the incidents appearing in Table 1 have only the most flimsy diagnostic value, as the descriptions of effects produced are vague or imprecise, for example, Nos 11, 21, 30, 35, 36 and 67. Only 29 can be analysed in any depth, namely Nos 15, 37, 38, 39, 40, 41, 42, 43, 44, 45, 46, 48, 49, 53, 54, 55, 62, 66, 68, 81, 83, 85, 86, 88, 89, 91, 92, 94 and 100. Of these only Nos 15, 38, 42, 48, 49, 53, 54, 55, 65, 81, 85 and 92 are really helpful diagnostically. Most of these appear in Table 2; those that do not are incidents not associated with blast—Nos 15, 48, 53 and 85.

Cases Nos 15 Perth N.J. and 48 Feyzin

Incidents Nos 15 (Photograph 1) and 48 are interesting in that they both involve sudden discharges of flammable material into an existing fire situation. In both cases the container failed catastrophically. Although the results were spectacular in the extreme and unfortunate in terms of consequent loss of life, no blast effects were reported from either incident. The failure of both containers allowed their contents to expand rapidly in three dimensions, thereby effectively maintaining a relatively narrow boundary of potential flammable vapour/air mixtures, into which the flame spread. In such circumstances flame speeds exceed mixing rates, and premixed flammable volumes are small, which results in the rapid development of an enshrouding flame envelope and only minor expansion effects. Thereafter no opportunity for air admixture in the presence of the flame can exist.

These two incidents are, of course, not unique—indeed, they are more the rule rather than the exception in any fire involving sealed containers of liquid flammables. They do, however, represent an extreme of scale which is rarely exceeded.

The behaviour following fire-induced catastrophic total failure is as anticipated; no conceptual difficulties arise. However, what is the situation following partial failure—again in a fire situation? Here there exist possibilities for "uni-axial" jets of flammable material to form. Precisely what happens depends upon the size, disposition and shape of the discharge orifice which is generated, and, more particularly, on the ratio of the orifice area to the volume(s) of the vessel(s) concerned (if interconnected).

From high-pressure equipment there is the distinct possibility that any accidental emission will flow at velocities greater than the appropriate flame blow-off velocity. Except in the presence of a considerable body of fire, therefore, the discharge *may* not ignite and burn stably. If it does not it will induce air into admixture by momentum transfer effects and so create possibly huge flammable volumes. The mechanics of the mixing process are well understood and have been verified experimentally[99]. What is not clear is whether an appreciable flammable volume can develop *in the presence of a flame*, although this has been postulated as an explanation for the particular case of Flixborough[21]. Clearly, the mass of flame in the vicinity of the flammable emission has relevance. (I am aware of one other UVCE in which the major pre-explosion emission occurred—or has been postulated to have occurred—in the presence of flame but the circumstances are far from clear.)

This, however, does not mean that such events have not occurred nor, furthermore that they are impossible. In this respect detailed consideration of the mechanics of momentum jet mixing are necessary. The most recent review of the subject and one which presents useful predictive equations is that of Marshall[100]. This work is cited fully in the Appendix and results obtained from it indicate that the size of flammable volume associated with a large aperture is vast and that not only will all but the largest sources become rapidly depleted, but also the initial—and transitory—momentum jet which is formed is likely to be rich or over-rich in the flammable constituent. Flixborough, therefore, may represent an event towards the lower end of the scale for which it is possible that a flammable cloud could develop in the presence of a potent source of ignition.

Case No 53 Gulf Ref. Co., Pa

The other vapour cloud ignitions which did not give blast and about which details are known are Nos 53 and 85. In these cases, as in the majority of UVCEs, the vapour cloud formed apparently in the absence of any potent ignition source. Incident 53[83] involved a mixture of cumene, benzene and propane escaping from a process pipework rupture. The process pressure was 35 atm and the ruptured pipe 150 mm diameter. The temperature of the process fluid is not known but, since the propane was described as gas in process and the benzene as liquid, it must have been between 85°C and 250°C. Accordingly, the discharge would have flashed, in part, to vapour immediately. Further, it is to be inferred from the description that it was further assisted to flash by impinging on a hot drum—indeed, the very drum from which the fluid was escaping.

It is not clear whether it is implied that the drum assisted by adding latent heat of vaporisation or merely by physically dispersing the efflux over a greater volume; the drum was, of course, lagged. What is clear is that the predominant direction of the emission was downwards. The rupture occurred at about 18.45 hours and the first fire alarm was given at 19.24 hours. It is presumed, therefore, that the escape lasted for about 40 minutes before ignition.

We have no information, however, on the actual size of the rupture nor the discharge capacity of the system. Photographs illustrating the reference indicate a drum of small size, possibly having a liquid capacity of 700 kg. It may be, therefore, that the initial violent impetus of the escape quickly became less fierce, that the rapid mixing—undoubtedly achieved initially—did not occur later in the discharge, and that a very low lying basically over-rich cloud of vapour resulted, presenting only a small volume within the flammable limits.

A wind is mentioned, but no force is ascribed to it. The source of ignition was identified as tube-still some 45 m down-wind of the source of vapour.

This incident, which is better documented than most, illustrates some of the difficulties attendant upon attempted analysis. The basic reference[83] was prepared by the Fire Commissioner. His chief concern in the article was to relate how an undoubtedly severe fire was tackled. The article does, however, contain more than the usual brief information on cause; nevertheless, important information in the context of this work is not included, some of which has already been indicated.

More particularly, however, details of the plant and process are missing. This is not a criticism of this Fire Officer or his men; very likely they would not have known what to ask for nor, probably, would they have understood it had it been provided. Furthermore, such information was not necessary for their task either as firefighters or advisors. Nevertheless, for the purposes of accurately defining the character of the discharge and its behaviour with time, details of the interconnected equipment and process fluid properties are necessary.

Case No 85 – Barcelona

The other incident in this category, No 85, involved a mixture comprising 70% dichloroethylene (DCE) and 30% vinyl chloride (VC). The explosive limits of this mixture will depend upon which isomers of DCE were principally involved: the CH_2=CCl_2 isomer is flammable[101] in the range 5.6% to 11.4% by volume, whereas the CHCl=CHCl isomers have the range 9.7% to 12.8%. VC is flammable in the range of mixture with air of 4% to 22% by volume. The widest limits of the mixture would therefore be, using Le Chatelier's rule[102], 4.8% to 14%. Chlorinated hydrocarbons have generally lower burning velocities than hydrocarbons.

In this incident a 300 mm diameter inlet to a centrifugal pump ruptured and discharged flashing liquid initially at 12 atm and 150°C. The vapour cloud developed for 75 seconds before ignition to cover an area approximately 150 m by 75 to 90 m. The flash of flame was brief (of the order of seconds) and intense, but only a very feeble pressure effect was created. The volume of the flammable cloud was estimated to have been 70 000 m^3 containing 20 m^3 (about 22 tonnes) of mixed organic chlorides comprising 10 000 m^3 of vapour and mist.

The form of the discharge is not clear here. Although a circumferential pipe burst would give a disc-like discharge, there is no information on the elevation or orientation of the ruptured pipe nor on what, if any, nearby pipework or vessels may have been in the vicinity to affect the mixing process. It is to be presumed that the pump was at ground level, since pumps usually are. A low-level cloud is therefore more likely. The proximity of the ground could have inhibited mixing.

It is interesting to note that the volumetric proportion of vapour and mist in the flammable cloud is, on the above figures, 14%–the probable rich limit. Flame propagation would have led to vaporisation of the mist and thus over-rich conditions. It is to be noted that the figures quoted are those prepared by the information source in this case.

It would seem that inadequate mixing with air could be the whole explanation for the failure of this cloud to create blast effects. The discharge rate and pressure drop through the point of rupture may have played a significant part in the poor dispersion—again the pressure drop *vs* flow characteristics of the rupture and associated plant are not known. Nevertheless, chlorinated hydrocarbons generally burn somewhat feebly. As before, therefore, one can find reasons for discrimination between these incidents and UVCEs. Whether these reasons are cogent will be considered later.

All the other cases selected for more detailed study are UVCEs. Three of these will be discussed in this section, Nos 38, 49 and 54; the remainder are deferred to the next section. Case No 38 is documented in some detail[76]. Of particular interest are the photographs to which reference for comparative purposes will be made later.

Case No 38 Doe Run

At Doe Run ethylene oxide was stored in twin horizontal 57-m^3 capacity tanks. Neither was full—the one which ultimately exploded contained about 25 m^3 of ethylene oxide. It would appear that an ethylene oxide decomposition was initiated by accidental ammonia contamination. Two things then happened: the rupture of the tank, followed by the ignition and explosion of the "vapour" cloud. The description[76] says: "it is the latter explosion which is believed responsible for the building damage". Clearly, there is some doubt about which explosion might have been responsible. Unfortunately no energy release can at present be assigned to the tank rupture as its design or test pressures are not known to me. The rupture energy release would not necessarily have been high, however, as high-pressure storage is not required for ethylene oxide.

It would seem, nevertheless, that the tank did rupture catastrophically and that virtually all the contents were released to atmosphere as a dispersed cloud. Ethylene oxide boils at about 10°C at 1 atm pressure. The quantity which flashed would naturally depend upon the temperature of the ethylene oxide at the moment of release and the ambient temperature—the former is unknown and unassessable.

Again according to the description, "structural damage was extremely heavy within 150 m, minor structural damage and extensive glass breakage occurred within 300 m, the effects of the shock wave were evident in a 5-km radius and the sound of the explosion was heard 60 km away". "Implosion" damage was observed within a radius of 200 to 300 m. The most spectacular damage, however, was to the storage tanks and their immediate surroundings. The ground had been cratered and the other storage tank, originally 3 m away from the one which ruptured, had been "thrown" 120 m. The ruptured tank broke into three major fragments, two of which were projected; one, a tank head, was thrown 800 m; the other, the tank body, 1200 m. The original dimensions of these tanks would appear to have been about 3 m diameter by 7.5 m long. The tank which was projected 120 m contained some 15 m^3 of ethylene oxide. Pipe bridges were displaced and the skirts of columns failed at welds. Two empty 57-m^3 tanks at about 300 m from the blast centre were pulled 300 to 450 mm *towards* the centre by implosion effects. One control room and a laboratory, respectively at 50 m and 75 m from the explosion centre, were completely wrecked. Other buildings were severely damaged within 140 m. Cooling towers at 180 m were stripped of cladding (Figure 15 of Troyan *et al*[76]), as was a steel-frame building (Figure 14 of the same article). Nevertheless, despite the loss of most emergency facilities, one fire pump could still be used and mains piped water was still available on the site.

Davenport[44] has presumably elicited by private communication the information (quoted in Tables 1 and 2) that the TNT equivalent was 18 000 kg, as insufficient information to calculate the figure is contained in the description and neither the TNT equivalent nor the yield are mentioned there.

Some features of this incident are exceptional even in the context of the UVCE behaviour–the projection of an entire tank and contents, probably weighing over 20 tonnes, 120 m, the very considerable projections of heavy fragments of a tank beyond 800 m and the creation of a crater at the explosion centre. These features suggest that behaviour was more akin to that of a condensed-phase explosive. The far flung fragments of the subject tank, despite its relatively low pressure of operation, suggests extremely rapid pressure rise. The formation of a crater is much more symptomatic of a condensed phase explosive detonation than of an aerial explosion. 18 000 kg of TNT, for example, would produce a crater about 13 to 15 m in diameter[23,103]. A reasonable estimate of the energy required to project the tank 120 m, however, corresponds only to that released by about 900 kg of TNT–the corresponding crater size would be about 6 m.

The total picture could therefore be as follows: decomposition of ethylene oxide was initiated and the process accelerated in the condensed phase, the local effect of which was to cause cratering and the generation of high velocity fragments and projectiles. Rupture of the tank caused rapid mixing of air and ethylene oxide and its decomposition products. These latter could well have been in a condition to spontaneously ignite. Alternatively, the shock wave from the initiating explosion, upon reflection from nearby surfaces–or otherwise–may have had a role in precipitating the UVCE.

The fact that the UVCE was apparently within milliseconds of rupture would appear to allow little time for mixing with air. Is, therefore, an alternative view tenable? Did *all* the damage arise as a result only of extremely rapid decomposition of ethylene oxide? If the products of decomposition are assumed to be carbon monoxide, hydrogen, methane and ethylene only[104], sufficient energy is available in this reaction, at 100% blast efficiency in TNT terms, for only *28%* of the TNT damage equivalent. The UVCE subsequent to rupture must therefore have supplied the balance of *72%* of the energy. It is unlikely, however, if the decomposition reaction is an acceptable explanation for the first phase of the explosion that this could have been solely in the vapour phase within the tank, as this reaction has an extremely low mass rate[5].

It is not known whether ethylene oxide can support a liquid-phase detonation–the evidence is equivocal–but it does decompose in a conventional manner with extreme rapidity. The decomposition reaction is dependent on temperature (the usual Arrhenius dependence) and, if venting of the tank was inadequate, it is "autocatalytic"–as the pressure rises due to increasing vapour pressure, so does the liquid temperature and hence the rate of decomposition[104,105]. The effect is well known and can be devastating[106].

Whilst the point has been made that liquid-phase detonation–that is, a stably propagating ultra-high-speed decomposition zone–of ethylene oxide would seem to be unlikely, simple theory[105] does suggest that equivalently high rates of decomposition could arise when unconfined, provided that confinement remains adequate at each stage as the decomposition autoaccelerates to this condition.

Clearly, understanding of what happened at Doe Run is imprecise. Even if decomposition of the liquid, as suggested above, was responsible to the extent of an equivalent energy release on rupture of the storage tank of 900 kg TNT, the pressure within the tank at the moment of bursting would have exceeded 680 atm. Plainly, with such a driving pressure the projection of the nearby storage tank and the fragments of the tank itself are understandable; moreover, extremely rapid mixing of residual ethylene oxide and its products with the surrounding atmosphere would have been assured.

Case No 49 – Raunheim

Case No 49 is simpler to analyse than the Doe Run incident, as the material involved—methane—is not susceptible to decomposition. In this case methane, surplus to requirements, was being vented to atmosphere. Liquid methane passed to a vaporiser having a total capacity (including that of the vent) of not more than 4 tonnes. The vaporiser was instrumented to control the internal liquid level. It was not entirely clear to those present what happened, but possibly the liquid level control failed and a slug of liquid methane was ejected from the vent. This would have been released at high level, as the vent rose vertically above the vaporiser to a height of 25 m. So much is conjecture—the following was observed.

The weather on the day in question was cold (–12°C) and still. Operators in the control room situated about 50 m from the vaporiser and vent observed a white mist moving slowly along the ground. Ignition occurred when the mist just reached the control room, although it is likely that the effective source of ignition was furnaces almost as far removed in virtually the opposite direction.

Structural damage was not severe and damage, by blast, to the plant was slight only. Most of the injuries, to about 75 persons—16 of whom were seriously injured, were occasioned by flying glass. Glass breakage was extensive at 400 m and slight at 1200 m. It is possible that building damage was slight only, even close to the "centre" of the blast, because failure of the large-area windows with which these buildings had been fitted, allowed rapid pressure equilibration. Naturally, this could only have happened if rates of pressure rise were much less than are associated with shock waves. Undoubtedly, there is much truth in this, since the movement of heavy objects within, for example, the control room is suggestive of tornado force winds which would accompany such pressure levelling.

It was considered unlikely that more than ½ tonne of liquid methane was involved from thermal and plant geometry considerations. This could form a flammable homogeneous cloud an average of one metre deep and 40 m radius. Although deductions from glass breakage have to be treated with caution (see later), it would seem that very close to the centre of the explosion ordinary pressure effects predominated, at intermediate distances pressure and wind effects coexisted, and in the far field shock-wave effects were dominant.

Case No 54 – Lake Charles

The Cities Service Oil Company explosion involved isobutylene. This formed a cloud which covered about 2 hectare before ignition occurred[84]. This cloud, when it burnt, according to the author of the description, detonated: "The plot of the overpressure destruction *vs* distance statistics fits exactly on known TNT detonation curves." The description continues: "Our units resembled ruins from the ravages of war. Two-storey buildings had been reduced to 4-ft piles of rubble, towers leaned in all directions, pipe runs looked like a blackened mass of giant waste spaghetti."

Despite these vivid descriptions there is, however, no evidence in the article which is characteristic of detonation. For instance, had there been a crater, I am sure it would have been mentioned; had there been severe fragmentation and projection, these also would have been mentioned—but they were not. Had the overpressure *vs* distance curves for TNT been followed exactly, as the article suggests, cratering and the other phenomenon should have been observed.

Accordingly, it is considered that what happened at Lake Charles was an ordinary UVCE. Much of the evidence relied on to produce the TNT equivalent was probably at medium or far distance from the centre.

Isobutylene in liquid form does not require a high storage pressure—less than 2.7 atm at 30°C. Thus it is improbable that sufficient dispersion was achieved at the point of escape to generate extensive flammable mixtures at source. The liquid would, of course, have flashed to give initially over-rich mixtures and later rapidly evaporating liquid in pool or droplet form at about -6°C. It would seem that aerial diffusion was the means of achieving flammable mixtures. A wind of "unfortunate" direction is mentioned[84], but its velocity is not. It was probably greater than 0.3 m s^{-1}, however, despite the apparent spread of flammable concentrations to only 240 m in 15 minutes[43].

The above again raises issues of fact, resolution of which would assist greatly. Interpretation based upon this published data would seem to suggest that an escape of moderate force only (10 kg s^{-1}) occurred into an almost quiescent atmosphere. If that had been so, it is difficult to see from where the impetus for mixing came; cold isobutylene vapour, even if wholly vaporised, would have had twice the density of the surrounding atmosphere. It will be clear that insufficient is known about this incident for deductions to be of much use. Less is known about most other incidents.

These examples raise many general issues. It would have been extremely helpful if any of them had been well documented. Until Pernis, Case No 55, however, none were sufficiently described in any published document. The large incidents since Pernis have progressively added to the picture, but all have had their shortcomings. As yet, there has been no incident about which certainty exists regarding all of the following factors:

(a) flammable material which escaped;
(b) the exact point from which the material issued to atmosphere;
(c) the orientation of the escape, its elevation and form;
(d) the volume and mass rate of escape of the material and its state—both as functions of time;
(e) the weather conditions at that time;
(f) the delay prior to ignition of the "vapour" cloud and its form at the moment of ignition;
(g) an accurate description of the mode of flame propagation and all perturbations and anomalies;

(h) other factors relating to cause, *ie*, fire observed *before* the vapour cloud developed or exploded;

(i) a comprehensive pattern of damage, or non-damage, correlated with distance and catalogued according to either the minimum pressure or impulse necessary to do the damage or the respective maxima which would just not cause damage.

In the following Section well documented incidents are considered. These comprise Case Nos 42, 55, 66, 81 and 92. Case Nos 55, 81 and 92–respectively Pernis, Flixborough and Beek–were the subject of official reports. Case No 42 was also well documented, but had more the character of an experiment and provides some salutary lessons in interpretation. Case No 66–Port Hudson– received close study from the US Bureau of Mines and was the subject of their Report of Investigations No RI 7752[91].

V
WELL DOCUMENTED UVCEs

These are particularly interesting, because they present not only opinions–they present *evidence*, some of which is susceptible to further or different interpretation. In what follows interpretation of the evidence will be shown to be inseparable from the initial assumptions made concerning the nature of the phenomenon. Techniques of diagnosis will be canvassed, as generally there is no agreement yet as to their universal application and validity.

Basically, the techniques comprise the selection of evidence capable of interpretation and then making the correct interpretation. As "evidence" in this context is usually rupture, irreversible distortion, or merely displacement, the "correct interpretation" aspect requires a multidisciplinary approach which is strong in fundamental mechanics. Whilst thermal effects can generally be ignored so far as the blast itself is concerned, unfortunately, most UVCEs are followed by severe fire in which subsequent fire damage occurs. Care must be taken to be certain that blast and fire damage are truly distinguished.

It is worth reiterating that diagnostics in UVCE studies are not straightforward. Again, care must be taken not to prejudge the type of event which has occurred and fit the evidence to that event. This error can be made more easily with UVCEs than perhaps with other phenomena, because the evidence will fit. This is largely due to the nature of the evidence (which merely embodies signs of the energy which has been expended), and the as yet unclear nature of the blast process. Caution has to be exercised in relating the displacement, distortion or rupture energy to the type of impulse which caused the movement or damage.

The Appendix contains details of the various factors involved–over-pressure, impulse and duration of positive and negative pressure phase, and "wind" effects–for both ideal blast waves and "ordinary" pressure waves of various profiles. Calculations are not presented in this Report but results, where appropriate, are to be found in the Appendix.

Case No 42 – Jackass Flats

This is one of two cases which gave rise to blast effects, albeit of a limited nature, where the discharge to atmosphere was wholly unimpared. "Partial unconfinement" was not an issue. Also, ignition occurred spontaneously, which in the discharge of a pure gas, hydrogen, was considered unlikely and was unexpected.

As was indicated earlier, this incident occurred during a large-scale experiment in which a rocket nozzle was being tested. In most of the tests in which the flow rate was greater than 0.5 kg s^{-1} it was the policy to ignite the hydrogen at the outset and continue to burn throughout the discharge. In these tests, however, which were conducted for acoustic purposes, the discharge was not to be ignited. Approximately 1000 kg of hydrogen was discharged from an initial pressure of 23 MPa through the converging-diverging nozzle to atmosphere over a period of about 30 seconds. The discharge rate was uniformly increased to the test figure of 55 kg s^{-1}, which was maintained for 10 seconds. Subsequently flow was uniformly reduced to zero. During the turn-down phase, at about 26 seconds, ignition occurred.

The test was fully instrumented and recorded. Infra-red film was used to record events cinematographically. Observers within 0.8 km detected no pressure wave, but those at 3.2 km heard the explosion, which rattled windows. A surge of ground-level dust was also observed. Damage close to the point of hydrogen discharge was widespread but superficial and indicated negative as well as positive pressure phases. Most of the damaged buildings (Figure 3) were lightly constructed, but were designed to withstand wind loadings of 160 km hr^{-1}. Wind loadings of this level create pressures of about 1.3 kPa. Damage indicated the *actual* strength of structures *designed* for that pressure had been exceeded at a distance of 83 m.

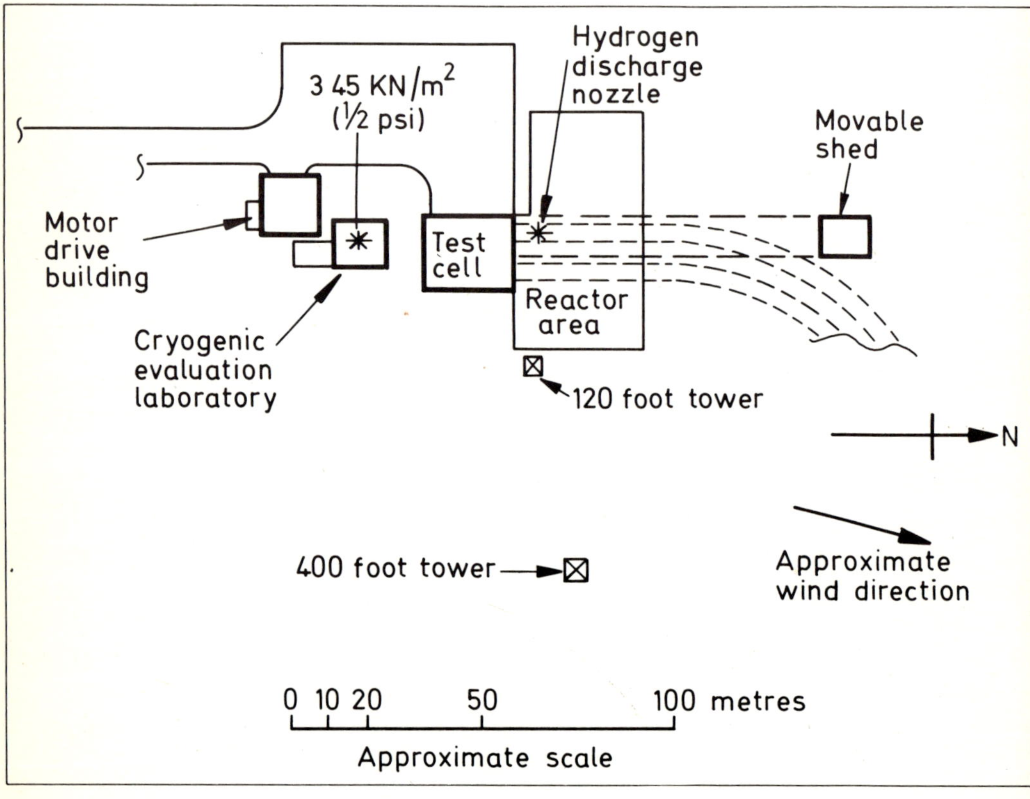

Figure 3 – *A plan of the site of the Jackass Flats incident – case No 42. (Taken from Reference 79 and reproduced by permission of Gordon & Breach Scientific Publishers)*

The infra-red film record disclosed that the burning cloud had a total height of about 45 m and an approximate diameter of 9 m. On a nearby tower (75 m distant) were wind measuring instruments situated vertically at heights up to 140 m. The instrument at 53 m height recorded a change in wind direction of 80° to show a wind from the direction of the test cell at the moment that a blast wave would have arrived. That wind velocity exceeded 29 m s^{-1} and broke the instrument. The wind at 77 m changed in direction by only 20° and the measured velocity increased by 25% to 2.5 m s^{-1} only. No effect was registered at 140 m.

The article[79] states that the dimension of the flammable zone as shown by the infra-red record indicated that 90 kg of hydrogen only was involved on the assumption of stoichiometric proportions. My calculations using the figures in the article (and their basis of calculation) suggests 80 kg rather than 90 kg at NTP with a stoichiometric ratio of 29.6%. Of course, what the film recorded was the *flame* and, furthermore, probably the furthest extent of the flame, since no reference is made to growth of the burning volume. Necessarily this flame-filled space implies the existence of flame temperatures within the volume and concomitant final expansion of the products from the pre-ignition state. The conclusion reached, therefore, is that the initial pre-ignition volume of flammable hydrogen/air mixtures was less by a factor of about eight than the one used in the calculation performed by the authors of the article. Not only does the quantity of hydrogen become less, on this basis–10 kg–but also the TNT equivalent yield of the explosion is increased to 8%.

Both the original calculations and those I have done above are crude. The use of Marshall's[100] correlations, in addition to enabling considerable refinement, adds insight. The Appendix sets out the relevant equations and their application to this problem. The physical scale of the jet system that was created during discharge has been determined from photographs and dimensions contained in the text[79]. Fundamentally, the scale of a turbulent momentum jet is prescribed by the discharge orifice diameter, which, by scaling, would appear to have been 0.61 m. Accordingly, the total height of the jet within the lower limit concentration envelope (corresponding to all hydrogen/air ratios in excess of the lean limit–4%) is calculated to be more than 1000 m. Further, the total volume of the flammable region of cloud (between the upper limit–75%–and the lower limit envelopes) is found by calculation to be about 3.2×10^6 m^3, and the corresponding total quantity of hydrogen in the fully developed momentum jet is found to be approximately 17 000 kg.

As the last figure is far more than the total quantity discharged, it will be appreciated that a fully developed flammable cloud may not have been established; although, it might be argued, the undoubted existence of standing shock waves at the nozzle renders the situation uncertain–especially in respect of mixing within a few diameters downstream of the plane of the nozzle.

This analysis has assumed no contribution from buoyancy. Reference 100 indicates that transition from jet to buoyant plume behaviour would probably lie between 840 and 1080 m along the axis of gas discharge. Turbulent momentum dilution effects would therefore seem to predominate. (Even here, however, some doubt remains, because of uncertainty with regard to the value to be ascribed to one of the parameters in the transition expression–in particular, the value of the orifice discharge velocity.)

On balance, therefore, the structure of the jet was turbulence- rather than buoyancy-dependent. But, to complete the picture, the buoyant plume—with a negligible discharge velocity—would have had a height, volume and hydrogen content, within flammable limits, respectively of approximately 340 m, 5.7×10^5 m^3, and 4200 kg.

It would seem that the actual zone of combustion was smaller than predicted by theory. If momentum jet theory is indeed appropriate, at the plane 45 m above the nozzle the central concentration would have been 52% and the radial distance of the boundary of 4% concentrations would have been at 5.7 m from the axis. A rough estimate of the amount of hydrogen present within the conical volume delineated by that circle, and the orifice as apex, indicates some 25 kg—clearly more in accord with the lower assessment of hydrogen involved.

It is valid to inquire why combustion apparently halted where it did. Central concentrations at the plane where extinction occurred spanned stoichiometric values. One explanation is that mixing rates were vastly enhanced beyond those predicted by conventional established theory. The nozzle velocity, assuming complete expansion, would have been about 2500 m s^{-1} (the article indicates half of this velocity: 4000 ft s^{-1}) at the maximum rate of discharge. Conventional adiabatic expansion calculations are not appropriate to diagnose what would have been the state of the hydrogen after expansion. This is because of the retrograde behaviour of hydrogen at ordinary temperatures; whereas most gases (and vapours) tend to chill, to the point even of condensing, on being decompressed, hydrogen conversely actually increases in temperature. The effect is modest but real—in this case a temperature rise of about 5°C would probably have occurred. Great temperature changes during simple mixing with air would therefore have been unlikely; enhanced rates of dilution, due solely to the creation of extreme thermal conditions, could therefore not have existed.

The only other powerful influences at work—and these undoubtedly did exist—were the standing shock waves. It is, however, difficult to see how these could actually have influenced *dilution* of the hydrogen by air, as these exist only by virtue of the high-velocity hydrogen stream. They are the boundaries through which the *hydrogen* decelerates and therefore *a priori* do not lie in and cannot affect mixing with air, since where they are situated no mixing can have occurred. Nevertheless, even if the shock waves were capable—even in principle—of having some effect, it will be obvious that the effect must be minute; because of the relatively huge mass of even small amounts of air inducted into the hydrogen stream, this will, by absorbing momentum, quickly decelerate the stream velocity to subsonic levels. Subsequent dilution through the flammable range would be by conventional subsonic means.

From this it will be clear that most of the 1000 kg of hydrogen should have been available to burn and yet the fact is that only about 10 kg did burn. It should also be clear that no explanation based upon enhanced rates of mixing can apparently explain this behaviour. This would remain a puzzle were it not for the explanations offered shortly and quantified in the Appendix.

What of the other factors involved? Again it is clear that observers in the near field were not conscious of noise, although it was heard at greater distances. Undoubtedly, however, moderate transient pressures were created and, equally certainly, air movement occurred. The fact that noise was not heard near to the explosion centre indicates that there the rate of change of pressure was outside the audible range. (In this regard it must be remembered that the ear responds most sensitively to rapid oscillations in pressure; thus the time interval between the rising pressure ramp and the subsequent descent is also important.)

Unfortunately, the durations of the pressure phases are not mentioned in the description, nor is it possible to gain more than an imprecise impression of their magnitudes. One interesting feature is the apparent lack of three-dimensional symmetry displayed by the pressure wave; at a distance of 75 m and an elevation of 53 m a change of "wind speed" from 2 m s^{-1} to considerably in excess of 30 m s^{-1} occurred, whereas, at the same distance, but only 24 m higher, the "wind" only increased from 2 to 2.5 m s^{-1}. Equivalent direction changes at the two elevations were also registered. It is difficult to interpret such polarised behaviour in what should have been a homogeneous medium away from the jet itself.

One factor, of course, is that those regions on the axis of the hydrogen jet—but beyond the limits of flame travel—would have manifest higher sound velocities. But it is difficult to understand why the effect should have been so pronounced if hydrogen concentrations beyond those limits were less than 4%. Another factor is axial symmetry of the volume involved in combustion, which might have been significant in the near field despite the relatively low height-to-diameter ratios of the burning region.

Pressures have been determined on the assumption that "wind" loadings in excess of the appropriate failure values were developed, that is, that pressures were wholly dynamic and without any element of shock overpressure. Many of the damaged structures appeared to have suffered a negative pressure phase, although caution is necessary here. Structures designed to withstand wind forces are so constructed that the wind pressure is resisted by members necessarily having the requisite strength to an *externally* operating force. If such a force arises as a result of an UVCE, the pressure transient may lead to internal pressurisation of the affected spaces through openings without causing structural damage at that time, only to cause subsequent damage when the pressure wave passes on and leaves the space so affected containing a high pressure relative to the surrounding atmosphere. The forces which are experienced then are of the opposite sense to those for which the structures are designed and failures can occur *apparently* as a result of a negative-pressure phase. Occasionally movement of an entire structure does show qualitatively that a negative pressure phase has occurred and the size of this can be calculated (at least in principle) from the amount of movement.

The article refers to a measured flame speed of 100 ft s^{-1} (30 m s^{-1}) in a vertical downward direction, which is roughly normal flame behaviour in a highly turbulent situation. If it is assumed that the flame speed was the same radially as downwards, then some 10 kg of hydrogen was consumed in about 0.15 second, liberating about 1.2×10^9 joule energy at an average rate of about 8×10^9 joule s^{-1} (8000 MW). If a uniform pressure profile is *assumed* (see the Appendix), having the form of the companion to the cycloid, then the maximum pressure attained at the centre of the explosion may be calculated to be about 17kPa (2.5 psig).

This assumes also that the observations of flame movement are real, that is, that what was observed was flame *propagation*, not the expansion of hot combustion products which had been already consumed by a much faster flame. It also assumes what is known not to have been the case: hemi-spherical symmetry. Explosion pressures 45 m from the centre of blast were found, from the observed damage, to have been about 3.45 kPa (0.5 psig). The theory outlined in the Appendix indicates that at that distance from the centre the maximum pressure in the travelling blast wave would have been about 3.12 kPa (0.45 psig).

Objects encountered by the blast wave in the near field would have suffered a crushing pressure lasting approximately 0.3 s and translational pressure-derived forces tending to dislodge radially outwards, followed by similar inward acting forces, both of 0.15 s duration approximately. Again, theoretically the pressure wave would have largely degenerated into a shock wave in rather less than 2 s or beyond 0.6 km from the centre.

Whilst the pressure wave propagates at sonic velocity, the air through which it travels, of course, does not. Nevertheless, it is displaced by moderate distances to the extent necessary to achieve the pressure distribution in the wave as it moves. This displacement causes a "wind", which moves with the pressure wave and exerts wind forces, which are distinct from the pressure forces.

Finally, the answer to the limited extent of flame travel in this case may not be enhanced mixing rate. It may instead have been due to the development of an atmospheric pressure distribution at the boundary of the jet having an inverse "pinch" effect (see the Appendix). Such a pressure distribution, it is emphasised, would arise *whatever form of blast wave occurs.* The effect would have been to cause the natural motion of the jet—already slowing down—to reverse and thus enable a barrier of air to intervene, thereby isolating the substantial volume of potentially flammable mixture, further downstream, from the advancing flame front. The proposition seems plausible theoretically and is in accord with the facts.

Case No 55 – Pernis

Pernis was the first severe explosion of the UVCE type to receive widespread attention. The only really comparable previous explosion in terms of scale and documentation was Doe Run. Because Doe Run involved ethylene oxide and was in any event a runaway decomposition, many of the issues it raised would appear to have been considered exceptional if not unique—unlikely to happen in other circumstances or, perhaps more pertinently, with other vapours.

Whilst evidence had clearly been accumulating (Table 1) indicating that this was not the case, Pernis, in particular, dramatically illustrated that such facile discrimination must be wrong. Not only were ordinary hydrocarbons involved, but their point of origin was so unexpected. The circumstances have been related in several documents[85, 86, 107].

The processes were operating normally at about 04.00 hours on the morning of 20 January 1968. Various plant operators and fitters *etc* were about their tasks. Shortly after that time, however, at least five men observed vapour spurting from a slops tank designated tank 402, situated in compound 24, itself located immediately to the west of the intersection of roadways 14 and 17 (Figure 4). Emission of vapour, it would appear, commenced at about 04.15 hours and ignition occurred at about

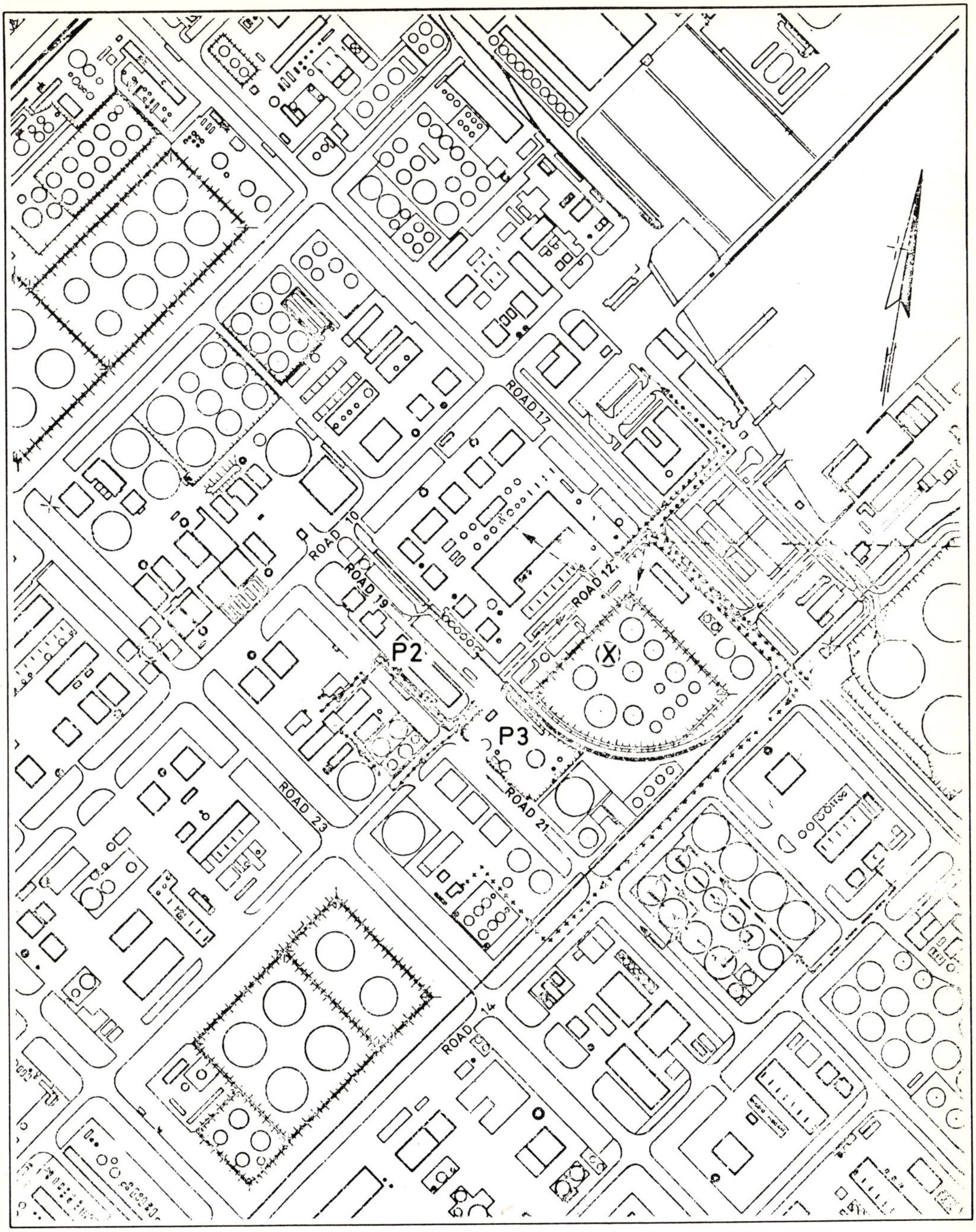

Figure 4 – *Plan of the area primarily affected by the Pernis incident – case No 55. The symbols P2 and P3 refer to Photographs 2 and 3 of this document. (This figure was taken from reference 85 – Shell copyright)*

04.23 hours. The weather at the time was almost calm; a very light variable wind was described.

On 19 January tank 402 was logged full of slops–the decanted oily waste accumulated from many sources in a refinery. Undoubtedly an appreciable quantity of water also found its way to this tank, but it seems likely that the major fraction of its 1575 m^3 volume was formed from hydrocarbon liquids. On 9 January there was a record in the log to the effect that steam heating had been connected to the internal heating coils of tank 402. This is often found to be necessary in the winter months to prevent partial solidification of waxes *etc.* It was thought that the explanation for the creation of vapour was that a layer of water-in-oil emulsion surrounding the heating coils suddenly separated at about 100°C after prolonged heating and that the oil thereby released caused boiling of lighter hydrocarbons and subsequent overflow of the tank. The resultant depressurisation then permitted the hot water layer to start boiling vigorously, thus generating, by heat exchange, copious quantities of hydrocarbon vapour.

Tank 402 had three "large" vents in its roof. It was unlikely, therefore, (but not impossible) that the roof was lifted by these events. Of the four eyewitnesses of vapour emission, three spoke of vapour spurting from the tank, but none mentioned a noise which might be construed as failure of the roof. The nature of these vents is not described and the roof was destroyed in the ensuing fire. Whether the emission was "vertical" (one or more roof vents) or "horizontal" (a split wall/roof seam) is not known.

The quantity of "vapour" emitted was clearly dependent upon the quantity of emulsion present initially and the effective boiling range and latent heat of the more volatile slops. It is not known if the tank was insulated, but it may well have been. If so, the effective evaporation potential of heat stored in the emulsion layer can be assessed by using figures already quoted[85]. The stored heat energy could have generated, on this basis, over 1300 tonnes of vapour. Plainly, this is an overestimate and there were probably many factors which would have lessened the quantity. Nevertheless, the potential was there and the 50 to 100 tonnes "officially" estimated cannot be regarded as unrealistic.

The mode of ignition of the vapour cloud was unknown and considered irrelevant[85]. However, various possibilities were canvassed in a technical appendix to the official report[85] prepared by M.C.H. Buschmann. Those considered were: electrical arcing, electrostatic sparks, hot furnaces and naked flames. Buschmann's conclusion is that it was not possible "with any certainty" to determine where the explosion was initiated. He states that the vapour cloud had become quite large by the time it ignited and the leading edge would have drifted approximately 100 m from the source. The meteorological report attached to the official report shows an *average* wind velocity of 1 m s^{-1}, changing in bearing between 04.00 hrs and 04.15 hrs from east to south-east. How it could be suggested that in eight minutes the leading edge of the cloud had moved only 100 m is not clear. Patently, a distance of 500 m is much more likely. Within that distance there were undoubtedly many controlled ignition sources (naked flames in heating devices).

Buschmann attempts to explain the UVCE in terms of conventional combustion experience. He cites detonation experience, in particular, before presenting the view that "a real detonation in space in a mixture of hydrocarbons with air must be ruled out". He does, however, discriminate; it "does not exclude the likelihood of transition from thermal explosion to a detonation-type effect in a large explosion cloud".

It would seem that Buschmann was responsible for determining the TNT yield of this explosion from the damage and assessing, *from the yield*, the quantity of hydrocarbons involved.

Few details of this damage are available to me. Certainly it was not instrumented, nor can what is available be translated into quantitative overpressure, blast or wind terms. The structural damage, Photograph 2, may be compared with later photographs of the incidents. The building in the foreground was some 120 m downwind of tank 402. The pipe-bridge shown in Photograph 3 was about 100 m upwind of tank 402 and was characteristically swept aside.

The overriding impression left by Pernis is the loss of an opportunity to learn as much about the phenomenon as would then have been possible. Without doubt, evidence of the type, since found elsewhere, must have occurred and yet no detailed technical appraisal appeared in the open scientific literature.

Case No 66 – Port Hudson

The only detailed account of this incident is that appearing in USBM Report of Investigations RI 7752[91]. The authors of that report had little doubt about the nature of the occurrence–"Detonation of a Flammable Cloud Following a Propane Pipeline Break." The US Bureau of Mines was invited to investigate this incident on behalf of the National Transportation Safety Board–the official body in the United States having responsibilities over the transport of materials by pipeline.

The incident occurred shortly after 22.00 hrs on 9 December 1970. As the locality was rural and sparsely populated, no one was immediately aware of what happened. A record of the occurrence was made, however, at the Villa Ridge pumping station about 15 miles (24 km) from Port Hudson, and "downstream" of that locality in the direction of propane flow. An "abnormality" was registered at 22.07 hrs. What the abnormality consisted of is not explained, but it was sufficiently abnormal for the pumps to shut down automatically. This had been accomplished by 22.10 hrs.

At 22.20 hrs there was a sudden increase in flow registered at the Rosebud pumping station, described as the "next upstream", presumably of Port Hudson as well as Villa Ridge, "indicating a substantial break in the line". The pipeline pressure at the time and place of the break was about 940 psig (6.5 MPa). Most of the pressure would have been hydraulic, as even at 4.4°C propane has a vapour pressure of 92.5 psig (0.64 MPa) only. Once the pumps shut down the excess pressure would probably have been quickly dissipated and the flow driving pressure difference would have fallen to about 90 psig (0.62 MPa) appropriate to the 1°C ambient temperature mentioned[91]. It is estimated that in the next 24 minutes 750 barrels (119 m^3) of liquid propane escaped out of the total lost during the incident, most of which burnt in the subsequent fire, namely 4538 barrels (720 m^3).

Most of those living nearby (see Figure 5) became aware of the escape because of the noise. Four families had evacuated their homes by 22.44 hrs and had met up on highway C to await and observe developments from ground about 100 ft (30 m) higher and somewhat more remote. What they could see was a white cloud developing in the valley below them. At 22.44 hrs the valley "lit up". "There was no observable period of flame propagation but a sudden flash as of lightning." Almost immediately there was a pressure pulse which knocked one witness over. "She was about 800 m from the centre of the cloud."

Various other immediate effects were noted, but the photographs best show the actual results. Photograph 4 shows damage to the house nearest the propane cloud (about 220 m from the "centre" of blast) and Photographs 5 and 6 show effects at progressively greater distance, namely 460 m and 610 m respectively. From their study of the damage the authors of the report, D.S. Burgess and M.G. Zabetakis, placed the 1 psig (6.9 kPa) and 0.5 psig (3.9 kPa) overpressure contours at radial distances from the "centre of blast" of 585 m and 1060 m respectively. Whereas the authors indicate no polarisation in effect up to 1 km, indeed up to 8 km, beyond that concentrations of damage resulted towards the north-east. Four such concentrations are referred to which, strangely, all lie in the one quadrant. The authors consider these anomalies to have been due to the refraction of pressure waves—so called caustic effects; they are not ascribed to the initial blast wave having an inherent direction.

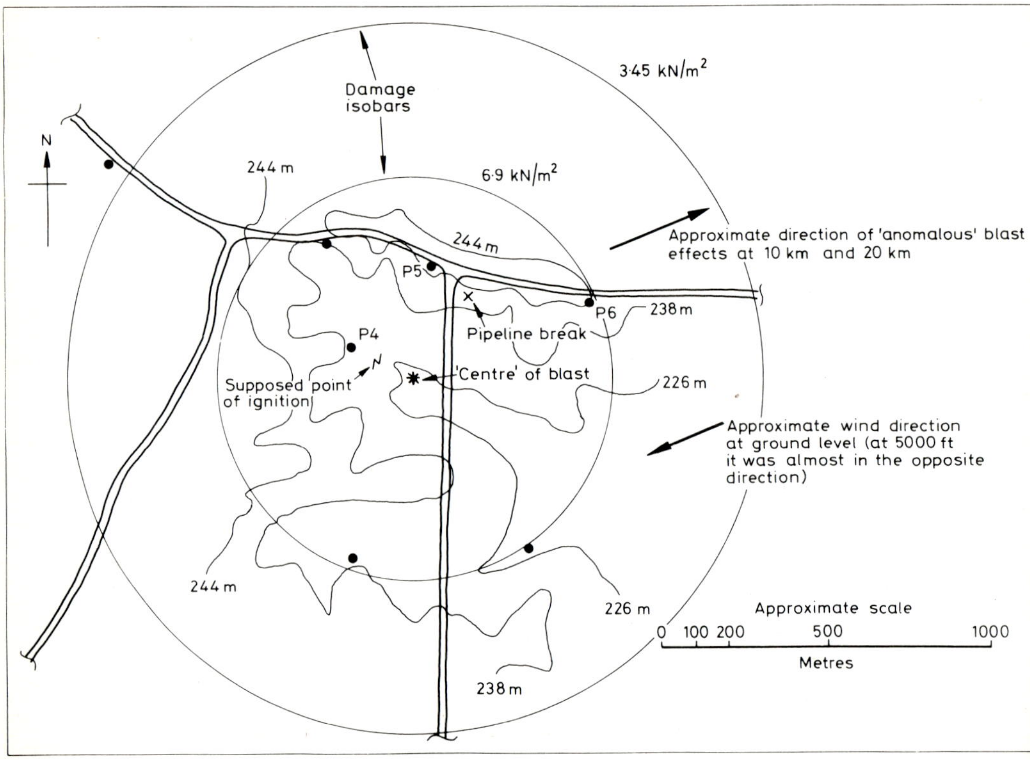

Figure 5 — *A plan of the area affected by the Port Hudson incident — case No 66. The symbols P4, P5 and P6 refer to Photographs 4, 5 and 6 in this text. (Figure prepared from reference 91 and reproduced by permission of the US Bureau of Mines)*

There are a number of apparent paradoxes in the report which ought to be highlighted even if their resolution is not possible now. The first is the most obvious manifestation of detonation to lay observers which would have been the bang, and yet no mention is made in the report of any nearby (or distant) eyewitness hearing a bang—nor does the report refer to a shock wave; only a "pressure pulse" is mentioned

The report also mentions persons within the 1 psi (6.9 kPa) "isobar", who should have been "in grave danger", but were unharmed.

The question immediately arises, from these comments—are we not rather discussing a pressure wave of the type canvassed in the Appendix and its concomitant effects? If we are, the data should fit the theory outlined in that section. That theory is idealised to a symmetrical and isotropic hemispherical system. Nevertheless, bearing these deficiencies in mind and using the information[91] that the release of 6.66×10^{11} calories (2.784×10^{12} J) and 6.9 kPa at 585 m and 3.9 kPa at 1060 m, then if $t_o = 0.78$ s, ϕ at 585 m is calculated to be (equation 30) 7.0 kPa and at 1060 m 2.2 kPa. The maximum explosion pressure predicted (equation 26) at the "centre of blast" is 286 kPa (41.5 psig) and the distance at which an identifiable shock front to the pressure wave would start to appear would be (equation 41) 138 m. These figures are interesting in that they indicate, superficially at least, agreement with the "blast wave" approach advocated herein. Rather more interesting, however, are their differences.

Figure 2 of the Report[91], from which the "isobars" are taken, presents an idealised and uniform pressure distribution, for which evidence is not presented and is unlikely to have existed. It is probable that the two isobars shown are no more accurate than ± 50%. Moreover, neither is it known whether the pressures referred to are side-on or reflected (the former has been assumed). That apart, however, and assuming for this purpose that the (side-on) overpressures stated were accurately assessed, it is probable that they fall within the explosive "static loading regime" (see Appendix) and that for comparison purposes the gradually applied pressure, concomitant with blast waves of the type posited here, should be doubled. The liberation of 2.784×10^{12} J in 0.35 s produces (equation 30) 15.9 kPa at 585 m and 4.9 kPa at 1060 m. These figures respectively over- and under-estimate double the isobar values quoted in the description. The error, however, does not all lie in this treatment since 6.9 kPa peak side-on pressure at 585 m corresponds to 68.8 tonne TNT, whereas half that pressure at 1060 m corresponds to 96.1 tonne TNT.

An hemispherical cloud of stoichiometric proportions of propane/air containing 70 000 kg of propane would extend radially for about 75 m. Central ignition implies an average flame speed commensurate with the above calculated figures of 96 to 215 m s^{-1}. Such flame speeds, whilst high, are *not* of detonation magnitude.

Ignition was considered ("speculate" is what the authors say) to have been from a refrigerator motor in an outhouse of one of the farms near the emission. The confinement afforded the flame whilst inside the outhouse is thought by the authors[91] to have given it the impetus to accelerate to "detonation". This is possibly the correct explanation; as the authors say, after 20 minutes of discharge, concentrations in the immediate vicinity of the pipeline break would probably have been too rich for electrostatic discharge or other, mechanical, effects there to have been effective sources of ignition.

Despite what has been said above, the probable situation was ignition of a fairly low lying cloud at its edge with the occurrence of "pinch" induced flame acceleration effects as suggested in the Appendix. If that was indeed the case, it is scarcely any surprise that an idealised explosion concept—whether TNT or "blast wave" based—should fail to fit the facts precisely.

Much of their description[91] deals with the development of the flammable cloud in terms of atmospheric dispersion theories. The result which the authors "find a bit difficult to believe" is that beyond the point of establishing a vapour cloud its size is then not influenced by further vapour emission: the problem being that the (theoretical) size of the cloud at that point is 3.1 to 4.3×10^4 m^3 of flammable propane/air mixture. The difficulty this raised was that the TNT equivalent of this volume, even on 100% efficiency basis, is less than 23 tonne.

Whilst it is clear that dispersion theory cannot be applied with accuracy to cold propane vapour issuing into the air at the bottom of a crater from a pipeline break, the discrepancy is, nevertheless, uncomfortably large. The TNT equivalent based on damage to nearby structures was found to lie in the range 44 to 227 tonnes. The authors considered 50 to 75 tonnes TNT represented the realistic range. Based on more remote window damage, 1.6 to 8 km, they found the equivalent to have been 45 tonnes TNT.

The efficiency of this explosion in TNT terms was based on the ratio of the energy release of 50 tonnes TNT to 70 tonnes of propane: namely, 7.5% of maximum theoretical yield. Whereas the authors consider the high value of 7.5% (compared to the estimates of Brasie and Simpson[107]–0.3 to 4%) to have been due to the direct initiation of detonation, there is no real attempt made to reconcile the paradox created by the theoretical appraisal: that the flammable volume at 100% efficiency could provide only 23 tonnes of TNT equivalency. The authors do, however, make the point that atmospheric diffusion coefficients for dense vapours are likely to extend the flammable volume. Upon using data for cold chlorine, they found a compensating reduction in cloud depth occurred so that, nevertheless, the predicted flammable volume actually changed only very slightly.

It is interesting to note the mention of evidence of hurricane force winds within the heat-affected area. Pressure gradients and pinch induced winds associated with the development of high flame speeds readily explain these effects.

Burgess' and Zabetakis' study of remote "focussed" damage is presented in detail in Section VIII.

Case No 81 – Flixborough

This is, by far, the best documented incident of all UVCEs. It is also the one about which most has been published[108–122]. Flixborough is significant as a turning point in the concern of engineers and others for the safe operation of plant containing, or otherwise associated with, large inventories of flammable materials. It was a large incident which caused damage in a populous area (Scunthorpe) and in a period of "social awareness". Whilst all the fatalities were employees of Nypro (UK) Ltd actually working on the plant, the subsequent Public Inquiry emphasised the general consciousness of what would have happened had this disaster occurred mid-week in a perhaps more typical industrial environment.

Naturally, the Public Inquiry[93] concentrated on the cause, or causes, of the explosion, and upon what and who was responsible. The difficult issue of why a vapour cloud should behave in this way was scarcely touched on. Many of the skilled investigators who later prepared reports (most of which have not been published) made assessments of the damage in terms of TNT equivalence. The assessments ranged from 15 to 45 tonnes TNT and were based upon effects produced at distances ranging from 100 m to several hundred kilometres (the ionosphere–see Photograph 7), including one at 19 km distance and 914 m height, being physically recorded on the barograph of a glider. This indicated a sudden pressure increase of about 360 Pa which, after allowance for the response characteristics of the barograph, was judged to correspond to what would have been caused by the detonation of 18 tonnes TNT.

There has never been dispute about the primary source of vapour for this explosion. The source was "obvious". Nevertheless, had Beek (the next incident to be considered in detail in this Section) occurred before Flixborough, perhaps the primary (in explosion terms) source of vapour might not have been considered so obvious. The source at Flixborough was "obvious" because of its size–two 700 mm diameter plant openings generated by the collapse of a temporary pipe. These allowed discharge of cyclohexane from a largely unrestricted 250-tonne-capacity system operating under saturation conditions at 150°C and about 1 MPa. What Beek introduces are queries regarding the necessary and sufficient size of a plant opening (or openings) to be commensurate with generating an explosive cloud–(a) big enough to precipitate the final UVCE, or (b) big enough on its own to create the whole extent of blast damage.

Causation was a keenly contested issue during the Inquiry[93], but that particular issue has no place here except possibly to consider afresh (a) above, which was a necessary factor in one of the strongly propounded theories. In brief, the proponents of this theory suggested that: an initial escape of cyclohexane occurred due to the localised rupture of an 8-inch (200-mm) pipe following a short period of intense and localised fire attack; the cyclohexane which escaped ignited as a cloud with sufficient force to blow down the temporary pipe running between and formerly closing the two 710-mm diameter openings mentioned above; and the vapour cloud which then rapidly formed gave rise to the extensively blast damaging UVCE.

That this theory was ever tenable, even in principle, arose, amongst other reasons, because of the observation of two deflections in the ionosphere (Photograph 7)–one of small magnitude followed by one of larger magnitude[123]. Both were much larger than could be explained by the simple release of pressure energy upon rupture. The earlier event was termed the "precursor" by some. The use of this word has rather special meaning in condensed phase explosive parlance, as is indicated in the Section on "Miscellaneous Issues" (Section VIII). It is considered that the "precursor", whatever else it might have been, could not have been generated by the main blast for the reason that no parallel medium (such as the ground) was available to transmit the pressure wave faster than the primary transmission medium–the atmosphere.

The ratio of the earlier to the later ionospheric effects was calculated to have been 1 : 50 in terms of blast energy. The present view[124] is that the energy equivalent of the main explosion was probably towards the lower end of the range 15–45 tonnes: namely 16±2 tonnes TNT. The energy equivalent of the "precursor" was therefore about $\frac{1}{3}$ tonne TNT or about 1.5×10^9 J. The same energy would be released by the combustion of about 35 kg of cyclohexane. If the combustion of this quantity of vapour occurred in stoichiometric proportions, the volume of air/vapour mixture involved would have been approximately 410 m^3, which corresponds to a sphere of radius 4.6 m.

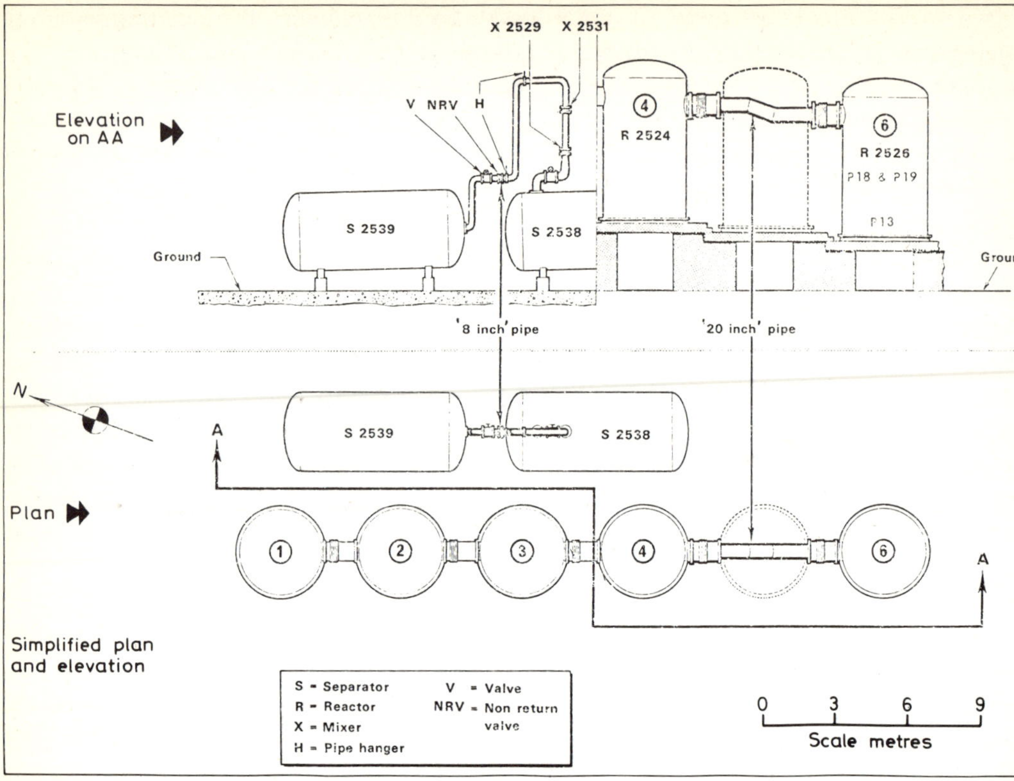

Figure 6 – *A plan and sectional elevation of the vessels immediately involved in the emission of cyclohexane in the Flixborough incident – case No 81. The notations P13, P18 and P19 refer to Photographs 13, 18 and 19 of this report. (Reproduced with the permission of the Controller of Her Majesty's Stationery Office)*

The actual (mean) distance between the point of emission, supposedly causing the vapour cloud (the 8-inch pipe rupture), and the temporary pipe was 13 m (Figure 6). Whilst it was not specifically put to test, it was the opinion of one of the experts[125], who provided a technical report, that the temporary pipe would have collapsed had it suffered an external distributed pressure force of $^1/_3$ bar or 33 kPa perpendicular to its axis. This corresponds to a pressure gradient at 13 m of about 163 kN/m^3 (after making due allowance for the diameter of the temporary pipe–20 inches or 0.51 m–and the skewness of the pipe to the (assumed) radial propagation of a blast wave). Applying the theory set out in the Appendix–in particular equation (32)–and using the values

$$\left(\frac{dp}{dr}\right)_{max} = 163 \text{ kN/m}^3 \text{ at } 13 \text{ m and } E = 1.5 \times 10^9 \text{ J}$$

then the volume of cyclohexane vapour would have had to be consumed in 0.017 s, which corresponds to a radial flame speed of 276 m s^{-1}.

A few comments are approriate: undoubtedly, turbulence in the issuing jet would have been high but the corresponding flame speed is far higher than can be attributed to turbulence alone. Considering the implications of equations (26) and (41), a steep-fronted pressure wave should have started to develop within a few centimetres of the centre of this "precursive" explosion. However, it would appear that no witness, with the possible exception of one of the laboratory staff, heard a *bang* at the time when these events should have been happening.

In view of these factors, it would seem to be unlikely that an explosion of the type considered in the Appendix occurred. Of course, it is possible to discriminate: the pipe may have fallen down at a lower pressure, the flammable cloud could not have been spherical and the dimensions of the flammable cloud would have been appreciable in comparison with the distance to the temporary pipe and the flammable cloud would have been subject to moderate to substantial confinement by columns, pipe racks and fin-fan coolers. Nevertheless, more cannot be made of the data nor do the apparently irreconcilable issues become compatible on any immediately obvious or straightforward basis.

Turning now to what undoubtedly was the major emission, liquid cyclohexane at saturation at 150°C and 1 MPa on decompression to 1 atmosphere will flash to liquid and vapour in such proportion that both are at their saturation condition, namely 80.7°C and 1 atmosphere. Equation (17) enables the mass ratio of vapour to liquid (μ) to be calculated (assuming constant physical properties over the pressure and temperature range involved). Using the values

$$c_L = 1.8 \times 10^3 \text{ J/kg°C}^{-1} \text{ and } \lambda = 6.74 \times 10^5 \text{ J/kg},$$

μ of the medium remaining in the reactor, is found to be

$$1 - \exp[-2.675 \times 10^{-3}(T_L - T)]$$

and, upon passing through the orifices to the open air, becomes 0.17.*

The reason is, of course, that with the flow of vapour and liquid from the system, internal pressures and associated temperatures are reduced—although not as far as if they were fully expanded, and the two remain linked by the vapour pressure characteristics of cyclohexane.

The "internal" μ, therefore, changes during the course of discharge, as well as changing as the two-phase mixture flows down the pressure gradient presented by the orifice. Probably the content of the vessels would have boiled violently as depressurisation progressed, in which case a more or less uniform mixture of phases would have resulted. It simplifies the argument if that is considered to have been the case. Equation (14) presents the criteria of minimum flow velocity through the orifice. For "internal" values of μ less than μ_{min}, c_{min} is the discharge velocity and the appropriate mass rate is obtained using equation (12) for ρ. For "internal" μ values equal to or greater than μ_{min} the discharge velocity and mass rate are calculated from equations (11) and (15) using the "internal" values of P, μ, ρ, ρ_v and ρ_L.

The initial mass flow from each of the two orifices could have been as high as 5900 kg s^{-1}, but would have fallen off rapidly, although to different extents. Equa-

* This value is sensitive to the values of c_L and λ and has been quoted as high as unity[124].

tion (15) indicates this on inspection, as μ increases from zero to infinity. Both the mass rate of the discharge and the residual mass in the system can be represented explicitly as functions of "internal" temperature or pressure, but not sufficiently simply for analytical treatment. Graphical or computed solution is possible therefore and in the case of the two discharge systems at Flixborough–initially containing 97.6 and 55.1 tonnes[126]–the quantities discharged from each in 20 s and the residual internal absolute pressures would have been 90.0 and 46.5 tonnes, and 2×10^5 Pa and 1.1×10^5 Pa, both respectively. Theory predicts almost total discharge in 30 s, which is confirmed by the evidence in the case of the smaller capacity system (see later).

It is probable, nevertheless, that the mass rates of discharge appropriate to the above were not attained (or, if so, were not sustained throughout) since the plant openings would have been above the equilibrium liquid level, except initially. This, therefore, is one source of error in the assessment. The other is the duration of the discharge period, which could not be better bracketed than 20 to 40 s. Obviously, the amount of cyclohexane in the cloud is equally open to error, saving that it was probably less than 100 tonnes. One recent estimate having authority[124], puts the cyclohexane content of the *flammable cloud* at 30 tonnes.

It will, of course, be recognised that entrainment, whilst it undoubtedly occurs, cannot increase the mass discharge rate beyond the maximum, as the entrained liquid will flash as it passes down the pressure gradient. In doing so it will compensate for its presence by reducing the pressure gradient and thus the flow. Lower discharge mass rates can occur, of course, merely by liquid failing to accompany vapour. The initial maximum discharge velocities would fluctuate between the following approximate limits according to the presence of liquid in the flow stream: 56 m s^{-1} (liquid present) and 195 m s^{-1} (liquid absent).

The discharge situation at Flixborough was complicated. The 0.71-m diameter orifices faced each other but were not coaxial; they were opposed at a distance of 6.13 m, the axes of the orifices being parallel but 0.356 m apart. Added to the fluctuating discharge, therefore, was the situation that both jets mutually interfered but in such a way that the zero-velocity locus, through which–at right angles to the principal discharge axes–a substantial volume of the emission would have passed was highly skew and probably not planar. Further, the vessels immediately behind both orifices were bulky and each of a size sufficient to intercept most of the undeflected portions of the jets. This was especially true of the vessel R 2524 (reactor 4), as this was situated at the higher level. Part of the discharge from reactor 4 would have cleared the top of reactor 6 (R 2526), possibly being deflected upwards as it did so.

Those parts of the discharges that moved into the south and west quadrants were largely uninterrupted in their subsequent movement, although progressively subject to the influence of the 12-knot, south-west wind, which was blowing at the time. North and eastwards, however, the emission encountered congested plant which probably hindered dispersion significantly. The picture which emerges during the few seconds before ignition is therefore an increasing volume of flammable cloud spreading in the west and south quadrants with less free movement of the vapour in the north and east quadrants, with the whole picture progressively polarised by the light prevailing wind. Again the reference[124] presents a plan of the probable situation, although there are some grounds for thinking that the cloud boundary defined in that reference corresponds to that of the final post-combustion condition[20].

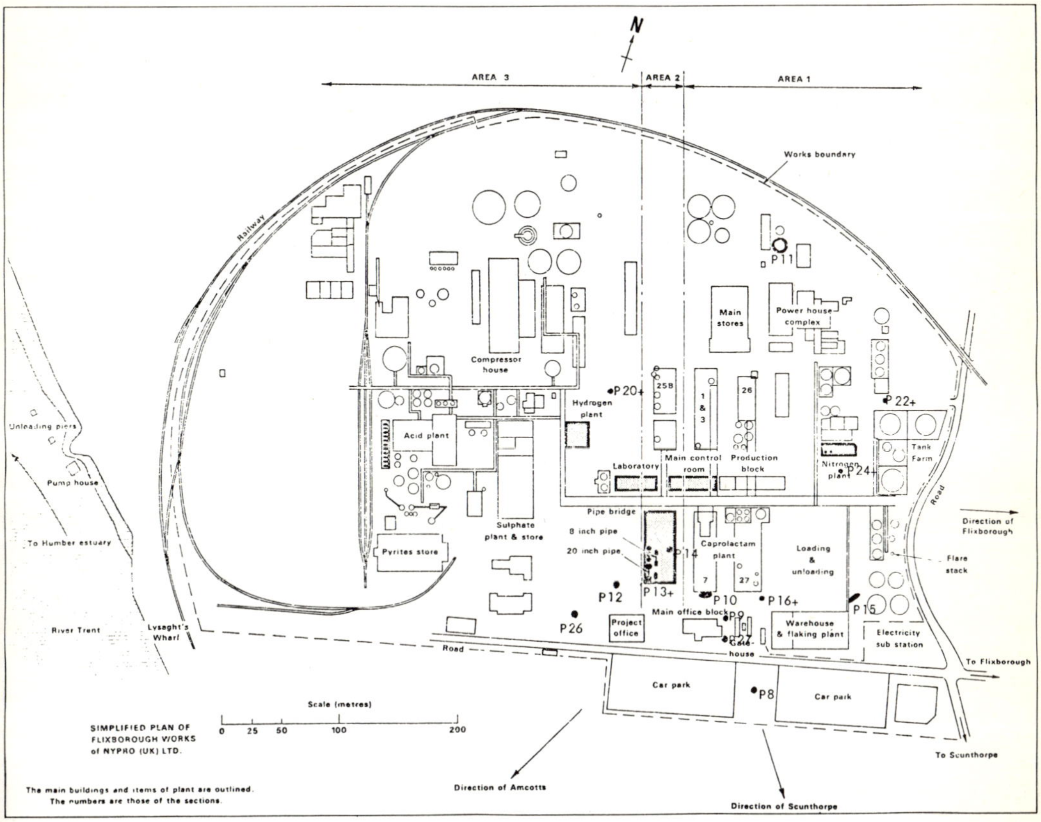

Figure 7 – *A plan of the factory site at Flixborough – case No 81. The symbols P8 through to P27 refer to photographs of the same numbers illustrating this report. The + signs after the numbers in some cases refer to other photographs at the same location. (Reproduced with the permission of the Controller of Her Majesty's Stationery Office)*

Nevertheless, the cloud so defined was "banana" or "boomerang" shaped in plan, extending from its eastern extremity westwards for 175 m and thence north-westwards for a further 165 m. The average width of the cloud over most of this extent was in excess of 100 m. The conditions of discharge precluded any appreciable vertical component to the cloud, which even at its lower limit would have been about 1.6% or more heavier than air. Not only in plan, therefore, but also in general three-dimensional shape, it would have appeared similar to a boomerang.

Sadee, Samuels and O'Brien[124] have made the point that it is "reasonable to expect" ignition of the cloud to have occurred at a boundary. The point they favour for ignition is the hydrogen plant (see Figure 7). Dr P.H. Kemp[127] conducted wind tunnel tests and indicated to the Inquiry's satisfaction that material from the orifice facing north could have travelled as far as the hydrogen plant.

However, whereas it might be thought that the cloud should, because of its density, lie close to the ground, Sadee, Samuels and O'Brien—by analysing the damage and equating it to what TNT would have produced—considered that the centre of blast was elevated above ground by a mean value of 45 m. This was a necessary artifice in order to enable fairly exact data matching with explosive especially close to the centre, as typical conventional explosive effects, such as cratering, were absent at ground level. It must be doubted whether any part of the actual flammable cloud attained 45 m before or during the phase of rapid energy release (that is, of course, on the basis of a single-event explanation of the incident).

Hitherto, almost all the assessments of damage following UVCEs have been related to overpressure alone and, in particular, peak reflected overpressure as typically appropriate to condensed phase explosive effects. There has been no separate study of dynamic (wind) pressure effects and no real endeavour to assess the actual duration of the pressure effects. It has apparently been assumed that these factors follow the same pattern in UVCEs as they do with explosives.

Only one appraisal of the effects of different pressure magnitudes, exerted for different periods, has appeared[21], which nevertheless adopted a simplified theoretical approach. This appraisal was restricted to analysing the pressure/duration effects commensurate with the crushing of the skirts of various columns. The details of this approach are presented in the Appendix. What would have been possible but has not been undertaken is a study of energetics, where deformation has occurred. This was possible with lamp posts, many of which preserved information concerning the direction of the blast wave (Photographs 8 and 9), being distorted through modest angles only, which enabled some confidence to be placed in calculations of the work done on them. Again, details of this analysis are set out in the Appendix.

One further aspect was the crushing of weak gas-filled containers, Photograph 10. It was recognised that these could provide little information from considerations of their strength to external pressure alone, but what appears to have been largely overlooked is that they all had "pneumatic" strength[21]. That is, where the container in question had an insignificant mechanical strength, it would, nevertheless, *not* crush indefinitely because of the compensating pressure rise (due to volume reduction) inside. Where this deformation happened in the explosion but the container did not rupture or was not otherwise vented the original dimensions were largely recovered (Photograph 11 of a large atmospheric pressure-water tank). The story was different, however, for vented or ruptured containers, for their original dimensions were only fractionally restored. It must be recognised that the existence of vents or points of rupture *does not* mean that for a given degree of crushing the crushing pressure is lower than would otherwise be the case. This is because during the crushing phase, air—or whatever the external medium is—flows into the container thus adding to the burden to be compressed. In fact, the converse is the case; for a given degree of crushing, the external pressure required to achieve it is higher, the larger the vent is.

One matter which conventional treatment does not allow is any estimate of the rate of pressure rise at or near the centre of an UVCE. The Appendix presents equations, which enable a mean value of this parameter to be evaluated—equation (26) and, by differentiation, equation (21) coupled with the value of $\dot{p}$ from equation (26) on the basis of a linearly rising pressure with time—not an essential but a convenient assumption. Three vessels are considered—C 2525, R 2525 and R 2526. The distance between R 2525 and R 2526 was 21.5 m and the distance from the point mid-way between them to C 2525 was some 38 m.

What R 2525 and R 2526 show (Photographs 12 and 13) is that crushing forces in excess of a nett pressure difference, between the inside and the outside, of 7.6×10^5 Pa (110 psi) occurred. The skirts of these vessels, and many others, were deformed in this quite characteristic way. I have reviewed this damage[21] and have concluded that no other cogent explanation exists. No other explanation has been offered in the open technical literature. The pressure necessary to crush the skirts of these vessels is significant—see the Appendix—in that it appears to be uniformly distributed, *ie*, occupying an appreciable volume, and *exceeds* measured pressures associated with deflagration in air in *closed and sealed* containers. Clearly some pressure intensification occurred which does not occur in less extensive situations.

If that were the whole story, it would be startling enough, but the skirts of these vessels—and indeed this was the case with all the crushed skirts—were provided with numerous apertures for ventilation, pipework and access. This meant that the rising external pressure was followed by a rising internal pressure always tending to counter the crushing force. The implications of this are set out in the Appendix in full, but, in summary, unless the pressure rose at a rate greater than a minimum value crushing could not take place. That minimum value for the reactors was 8.2×10^6 Pa s^{-1}. The time this would have taken would have been 0.185 s and the free atmospheric pressure at crushing would have risen to 1.518×10^6 Pa (220 psig). Lower atmospheric pressures (but still capable of crushing) would have had to be achieved rather more rapidly; 1.039×10^6 Pa (151 psig) could crush these vessels provided the pressure rose at 1.039×10^7 Pa s^{-1} or faster.

These observations apply equally to all the relatively strong crushed volumes including C 2525 (Photograph 14). The quantitative differences reflect their structural differences, but the damage can, nevertheless, be related, in principle, to their relative distances from an explosion centre. The three points represented by these three vessels hardly qualify to construct a blast-wave pattern. Suffice it, however, that they illustrate the principle rather than offer a definitive picture.

If it is assumed that the flammable cloud contained 30 tonnes of cyclohexane in premixed stoichiometric proportions with air, the combustion energy available for release was 1.314×10^{12} J. Now it is necessary to assume that central pressures rose faster than 8.2×10^6 Pa s^{-1}; otherwise, the crushing damage extant could never have occurred. Such an assumption requires, on the basis of the theory in the Appendix, that the rapid release of energy was accomplished in a mean time of 0.3 s (equation 26). This, however, is longer than necessary to cause crushing (0.185 s is the necessary time). The higher rate of pressure rise of 1.755×10^7 Pa s^{-1}, which will achieve the crushing pressure in 0.05 s, is itself only achieved when combustion is sufficiently rapid to consume the flammable cloud in 0.25 s.

The corollaries of these figures are interesting, bearing in mind that they are based upon the necessity to do the damage which occurred, namely: 8.2×10^6 Pa s^{-1} acting for 0.3 s would have given pressures at the centre of the flammable cloud of 2.5×10^6 Pa (357 psig); or 1.755×10^7 Pa s^{-1} operating for 0.25 s would have created pressures as high as 4.4×10^6 Pa (637 psig).

What does this mean in terms of flame speed? Approaching the problem on the basis of an hemispherical cloud, 8.2×10^6 Pa s^{-1} means an average radial flame speed—from a central source of ignition—of 184 m s^{-1} and, similarly, 1.755×10^7 Pa s^{-1} means 221 m s^{-1}. Both are high, but subsonic. Nevertheless, they are averages. Equation (58) gives the respective maximum velocities at the conclusion of combustion as 254 m s^{-1} and 294 m s^{-1}. Again high, but subsonic. But this is an over-simplification; if it is accepted that the rates of pressure rise calculated to do the damage are the minimum necessary and if the total energy available has been realistically assessed (it could, it appears, however, have been three or four times as much), having regard to the practical disposition of the flammable cloud, flame velocities of a supersonic nature are implied, but not necessarily detonation.

The values of rate of pressure rise and time to use in the crushing equations, if correlation is to be attempted with blast waves of the form described here, are for the former the mean value, either

$$\frac{p}{t_o} \text{ or } \frac{\phi}{t_o}, \text{ and } t_o \geqslant \Delta t.$$

For any particular case these parameters can be generated from the graph shown in Figure 8 and the values of p or ϕ, t_o and r (or τ) which result used to pin-point the

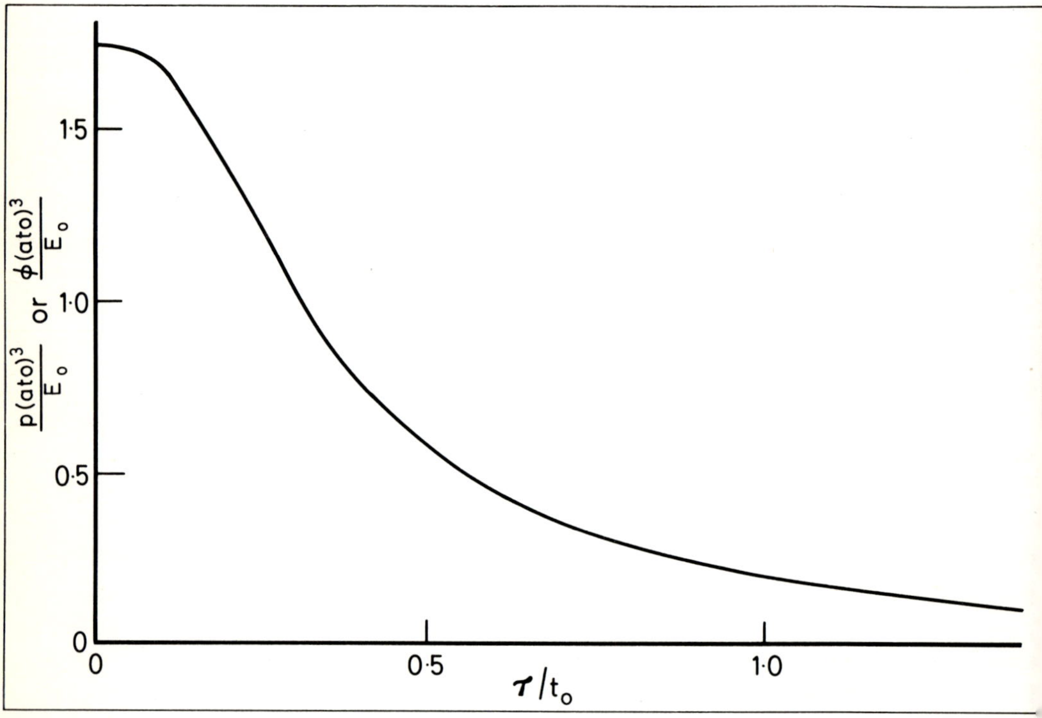

Figure 8 – *This represents the variation of the parameters* $\frac{p(at_0)^3}{E_0}$ *and* $\frac{\phi(at_0)^3}{E_0}$ *as a function of* τ/t_o. *The figure is constructed from equations (28) and (29) of the Appendix and relates the maximum pressure at any time (or distance) from the "centre" of explosion*

centre of blast. As there are three independent variables here (not even allowing for uncertainty in the energy release, E_o) and three damaged vessels it may appear to be possible, in principle, to locate the centre of blast. In fact, that is not the case, as one of the independent variables, Δt, is indeterminate (being $\leqslant t_o$) and the other, and possibly more difficult factor, is that the calculation assumes that crushing pressures were only just exceeded. Whereas that might have been true in the cases of the reactors (Photographs 12 and 13), the appearance of the skirt of column C 2525 indicates a considerable excess of pressure over the minimum crushing value.

What might assist in this case would be some knowledge of the extent to which the volume within the skirt had been reduced. Unfortunately, collapse had distorted the actual lobar deformation so no assessment was possible. The assessment would not be easy in any event, however, because the strength to external pressure is high only when true circularity exists. Moderate deformation only will lead to a rapid reduction in resistance to pressure and the difficulty in interpretation arises when, or if, the reduction in volume is assessed, what additional pressure factor should be applied to allow for the residual mechanical strength of skirt material–presumably beyond a certain degree of crushing its strength begins to rise again.

The issue is not so clear cut as with weak containers, *eg* Photograph 15 of the road tanker. Here the pressure resistance of the constructional material can be totally ignored. Again the difficulties exist of assessing the degree of crushing and, once that is assessed, making allowance for the presence of condensible vapours. For this particular vehicle the crushing pressure has been estimated[21] to have been between 3.4×10^5 Pa and 10^6 Pa. Here again the scale of events is highlighted; from the reactors to the remains of this vehicle was a distance of 180 m.

Were these isolated pockets of high pressure? It seems unlikely. Virtually on the straight line between them and 100 m from the reactors was the drain cover depicted in Photographs 16 and 17. This was formed from cast iron and was 28.6 mm thick and 1 m by 0.6 m in area. The cover had broken into four roughly equally sized fragments with the fracture edges parallel to and approximately half way from the sides. Impact damage does not seem to have been responsible. Failure theory suggests that pressures well in excess of 1 MPa had caused this damage.

At least one other feature of the damage indicates the development of very high pressures. This was deformation which occurred actually *within* reactor No 6. The fact that damage could have happened within reactor 6 should cause no surprise in view of the likelihood, considered above, that that reactor should have exhausted its internal pressure within 20 to 30 s. Photographs 18 and 19 illustrate the damage. Figure 9 is helpful in understanding the situation. The baffle plate over the inlet to that reactor was blown off and impacted the central stirrer shaft, which itself bent, as a result, through 15° over a length of about 0.6 m. The stirrer shaft was fabricated from stainless steel tube of 150 mm overall diameter and 25 mm wall thickness.

The load which will just cause yield in the outermost fibres of the stirrer shaft when applied 0.6 m from the point of cantileverage is approximately 1.1 MN (assuming a yield stress of 414 MPa). The work done by this load is greater than the result obtained merely by multiplying the load and the distance of deflection because of "springback" due to the release of residual elastic strain and other factors; nevertheless, the work done was a minimum of 8.6×10^4 J.

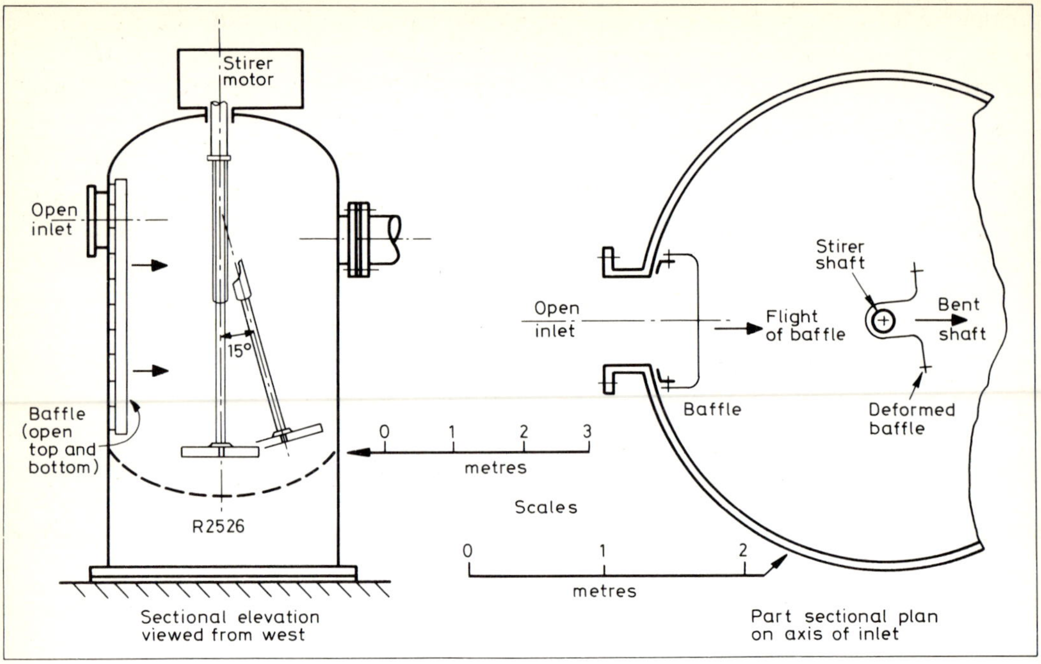

Figure 9 – *A diagram indicating the damage found within a reactor at Flixborough, the reactor concerned being No 6 (R 2526 – shown in Figure 6). The damage consisted of a bent stirrer shaft and a deformed baffle plate, the bending of the stirrer shaft being occasioned by the shearing of the baffle securing bolts and its subsequent flight into impact with the stirrer shaft, all as a result of the development of pressure outside the reactor. Photographs 18 and 19 illustrate this damage*

This work was done by the baffle plate firstly shearing all 14 of its 9.5 mm retaining bolts and then accelerating to gain sufficient energy to bend the shaft on impact with it. The baffle plate was, of course, a bluff body and, equally clearly, the reactor was not evacuated; but, despite these unquantifiable energy dissipating factors and assuming energy was conserved at each stage of the process, then an excess pressure, over the residual internal pressure in the reactor, of at least 0.64 MPa (93 psi) probably existed.

One other factor of diagnostic value, both because it indicates where the "focus" of the blast wave was when deformation occurred and because it provides evidence of energy expended, is deformation of lamp standards. Those at Flixborough were supplied by the British Steel Corporation to two specifications–columns: 30–32 (September 1969) and 5-02 (May 1970). They were formed from steel tube. Each type had a short length at the bottom of larger diameter in which were situated the lamp controls. The two were differentiated on the site (after the explosion) by the different types of lamp fixing at the top.

Photographs 20–25 inclusive are in pairs, each pair showing the same lamp-post in the plane of deflection and at right angles to that plane. In the background of the former photographs may be seen items of plant recognisably close to the centre of blast. In the Appendix is presented an approach which enables information to be obtained from this damage. In all cases, where deformation occurred, the fact that it happened meant that a local pressure gradient in excess of a certain critical value (the yield value) must have existed. Although this pressure gradient probably exerted a uniform force over the whole height of the lamp-post, conventional cantilever loading theory indicates that the greatest bending moment would arise at the root of the cantilever. It was there that bending had occurred in the majority of cases, the root of the cantilever being at the change of cross-section.

As with the stirrer in reactor No 6, the strain beyond the elastic limit is approximately calculable; therefore, the work done by the pressure wave may be roughly ascertained. The calculation is approximate only, for the reasons indicated already for the stirrer. In this case, however, there are additional possibly significant complications; during deformation the pressure gradient changes; during deformation the "profile" presented by the lamp-posts to the pressure wave foreshortens; if deformation is rapid, inertial effects due to the rotational motion of the lamp-post may be large; the reverse pressure gradient associated with the rear of the pressure wave may restore the deformed shape somewhat.

In most instances deformation was in one plane, although near the centre this did not always occur—see Photograph 26. Further, in most instances deformation apparently occurred in one stage, although again near the centre this was not so—see Photographs 9 and 27. A valid question with regard to those lamp-posts, which were deformed in separate stages and/or in different planes is why, if the greatest bending moment was at the root of the cantilever, did not all the bending occur there? One explanation might have been the simultaneous existence of contrary pressure gradients at different elevations; another is that after considerable yield over one region the metal there displays work hardening so that the likely point of subsequent bending moves a corresponding distance along the lamp-post.

In the Appendix I have calculated the energy required to achieve deflection for a particular case. This parameter is, however, difficult to make further use of. As a minimum energy level (expressed in terms of the minimum necessary pressure gradient to do the damage) is required before any permanent deformation can occur, the calculated energy is the excess over the maximum energy level which will *just not* cause deflection. Although I have not taken this aspect of damage further, undoubtedly it could be extended possibly to some effect.

Flixborough has enabled diagnostics, as applied to events of this type, to be extended. I have presented these in detail to illustrate the approach I have evolved. Again, whether this approach is justified must be judged only on its merits; its wide applicability and its consistency.

Case No 92 – Beek

On Friday 7 November 1975 an explosion took place in the vicinity of the naphtha cracker II installation on the site of Dutch State Mines, Beek. The explosion was immediately followed by a fire in and around the installation. 14 persons were killed, of whom six were working in the controlroom building and eight were outside this building in the open air. 104 persons were injured. The installation and a nearby tank farm were destroyed; outside the plant area there was extensive glass damage. The installation was in a start-up condition, after the plant had been out of use for a long time. The explosion was caused by the escape of a C_3 fraction from one or more ruptures in the plant used for its extraction. About 5500 kg of hydrocarbons escaped, which mixed with the air forming an explosive cloud. This cloud moved through the installation and was apparently ignited by one of the cracker furnaces.

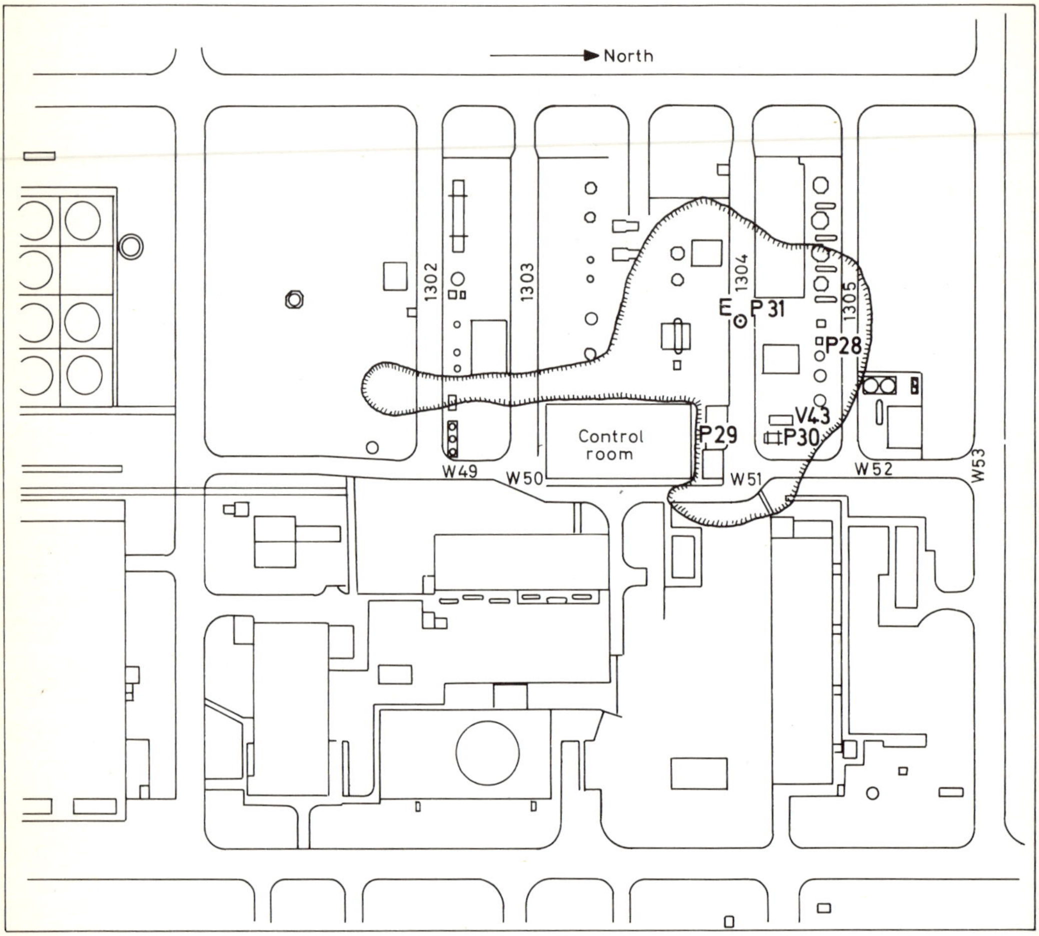

Figure 10 – *A plan of the factory which was the subject of the Beek incident – case No 92. The symbols P28 to P31 inclusive refer to photographs illustrating this report. (Taken from reference 97 by permission of the Netherlands Ministry of Social Affairs)*

The Court of Inquiry was informed in a report prepared by the Rijksverdedigings-organisatie TNO that the damage which resulted was similar to the damage which would have occurred had 2200 kg of TNT detonated. It was further their opinion that of the 5500 kg hydrocarbons involved, only 800 kg exploded. On this basis the yield becomes 28% rather than the 4% which is quoted in Tables 1 and 2. The basis for the view that "800 kg exploded" is obscure, but may be related to the size of the cloud as described by eye-witnesses. This was presented in the Official Report as extending 30, 45, 30 and 90 m respectively north, south, east and west. It was not given a thickness although it was originally observed 3 to 3.5 m above ground level.

Figures 3 and 12 of the Official Report[97] indicate the size and shape of the cloud in plan (although they differ). Figure 12 forms Figure 10 of this document.

As with Flixborough two hypotheses on cause existed in the early stages of the inquiry: a process fault, which led to excessive pressures, or the explosive decomposition of organic peroxides. The technical study favoured the former explanation. The process was, as has been stated, under start-up and had not yet reached the desired operating condition. Specifically, the temperature and composition of the liquid stream to V-52 feed accumulator had not attained 17.5 atm and 56°C. It has therefore been assumed that the feed from the deethaniser column, C-08, to V-52 was too cold so that lighter fractions entered the vessel. The reboiler H-58A of the depropaniser C-11A was commissioned just before the explosion which caused process stream temperatures and pressures to rise very significantly in a short time. Also the working of the demethanator C-06 was upset by the trip of a refrigerating compressor, K-03.

Only two of the five cracks, ruptures or breaks found after the blast were thought by RVO-TNO to have been capable of discharging 5500 kg of propylene on the basis of calculation. One of these was just under, and upstream of, the safety valve of V-52 (under SV-26 at B1–Figure 11), Photograph 28, and the other was upstream of the safety valve on the reboiler to column C-11A (under SV-27A at B4–Figure 11).

On balance the general tenor of the Official Report favours the rupture of the pipe between V-52 and its safety valve SV-26. This pipe was about 50 mm in diameter and the discharge lasted rather less than 2 minutes before ignition.

The Report describes the damage and makes the point that the most severe damage occurred where the greatest concentrations of equipment–within the flammable cloud–were to be found, such as pipe-tracks, *etc.* The investigators "were able to show" that the focus of blast was approximately at the centre of the cloud as indicated by the "E" in Figure 10. The control room–Photograph 29–located as shown in Figure 10 (Meetkamer 1 : G13-04) was about 20 m from the centre of the blast. The calculated excess pressure loading it suffered was 0.02 to 0.03 MPa.

Two additional items of plant damage received detailed study with the aim of evaluating explosion pressures. One was the cold blow-down drum V-43 (Photograph 30) and the other a cantilevered reinforced concrete platform (Photograph 31). The estimated pressure to which V-43 was subject was 0.1 MPa. The platform was dealt with differently, however, in that its failure was related to the *minimum* necessary impulse level to do the damage: 3500 Pa s.

These detailed in-plant damage studies are, to my knowledge, the first, with the obvious exception of Flixborough, to have been undertaken. RVO-TNO considered in detail the course of combustion of the flammable cloud. After ignition at the edge of the cloud the flame behaved "normally" for a short distance before accelerating to high speed. The reasons for acceleration are stated to be "expansion ahead of the flame" and "turbulence effects . . . . . favoured by the presence of various obstacles".

RVO-TNO used the theory of Kuhl, Kamel and Oppenheim[128] to establish that 0.1 MPa corresponded to a flame speed of 250 m s^{-1}. They also "established" from

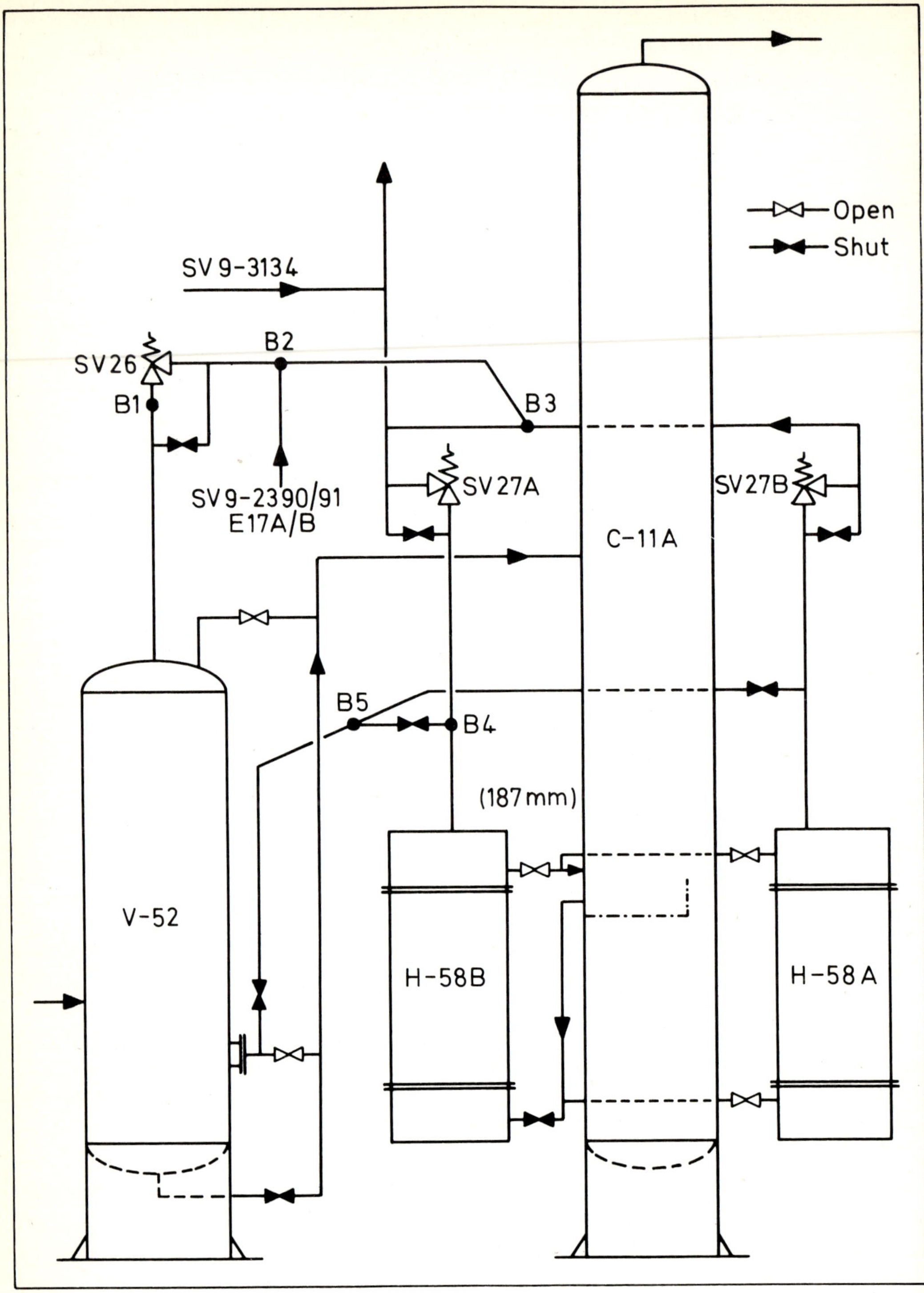

Figure 11 – *A process flow sheet of the section of plant which gave rise to the emission of propylene at Beek. (Taken from reference 97 by permission of the Netherlands Ministry of Social Affairs)*

conventional blast theory[41] that the "positive-phase overpressure" duration was 0.075 s. They considered, however, that the equivalent TNT effect could not be achieved by a single charge of TNT of 2200 kg detonated at one point. Their view was that the "best fit" scheme of equivalence was that involving five separate charges of TNT each of 440 kg placed "at the centre of the street between W-51 and W-52(?)"—see Figure 6—"with the first charge at 30 m from W-51 and the following charges at 5 m further in the direction of W-52(?)." Clearly reference to Figure 10 will confirm that W-52 is in error and should have been W-55. Plainly W-52 is in the wrong direction and is, in any event, only 25 m from W-51.

RVO-TNO state (implicitly) that they would have anticipated similar overpressure and impulse effects had these charges been detonated from the W-51 end at 0.020-s intervals (the time taken by a flame travelling at 250 m s^{-1} to cover 5 m). Whilst it is clear that the detonation of five charges over a period of 0.080 s cannot give an overpressure duration of 0.075 s, it may be their equivalent effect is just that. At this stage it will be obvious that the discussion is becoming remote from the problem; discussing as we are the equivalence of equivalents. An impulse of 3500 Pa s delivered in 0.075 s would correspond to a mean pressure during that period of about 0.047 MPa. In fact the impulse must have been delivered in a much shorter time, because of the pressure equalisation, which would have occurred on the opposite face of thin structures such as the platform.

The platform was only 1.175 m wide and 0.24 m thick, therefore a ramp shaped pressure discontinuity from below travelling, as it must, at sonic velocity would begin to generate contrary forces above the platform opposing movement—unless this was already under way—within 0.00075 s. Such an argument suggests that the mean pressure exerted must have been 4.7 MPa. Of course, this is an over-simplification, but probably brackets the pressure range (two orders of magnitude) capable of causing this damage. Unfortunately the civil engineer's evaluation of the failure pressure is not presented, nor is it calculable with accuracy from the figure in the Official Report (Tek 14), but from this figure a rough assessment of the minimum failure pressure gives 0.3 MPa.

This is a high figure and corresponds to a mean impulse duration of 0.01 to 0.02 s depending upon the pressure wave profile (assuming always a steep leading edge).

The alternative view is that 0.3 MPa was the nett pressure exerted by a pressure gradient of 1.25 MPa m^{-1} on an object 0.24 m thick. As this platform was within metres of the "focus" of the explosion—and if it is assumed that this pressure gradient was the maximum that existed—then from equation (22) in the Appendix $\dot{p}_c$ is calculated as 2.63×10^8 Pa s^{-1}. If it is further assumed—contrary to the RVO-TNO view—that all the propylene contributed to the blast, the energy released would have been 2.52×10^{11} J. To have generated the rate of pressure rise calculated above from the release of this energy, means from equation (26) that rapid combustion must have been completed within 0.08 s.

Assuming hemispherical symmetry and central ignition to which the theory may be appropriate—but, it is recognised, the physical situation was not—then the necessary mean flame speed to achieve this rate of release of energy may be calculated as 367 m s^{-1} (5500 kg propylene in 5% volume admixture with air at 0% and 1 atmosphere). This corresponds to a final flame speed—from equation (58)—of 490 m s^{-1}.

Naturally the deductions made here and by RVO-TNO depend crucially on the accuracy of the pressure of failure of the platform. A completely different result is obtained if the failure pressure, 0.047 MPa (corresponding to 3500 Pa s over s), is used. In this case the requisite mean flame velocity, using the approach developed herein, is 230 m s^{-1}, or a final flame speed of 307 m s^{-1}.

Damage outside the plant was also surveyed and was found to be largely only broken windows. It is not clear how this damage has been used in the assessment of TNT equivalence–if at all. The Official Report states that window breakage was high due mainly to the thin quality of the glass used in the older houses.

General Comments

All of the cases cited in this section, and many others, have already been appraised for their explosive force by comparing the damage which occurred in each case with that produced by the detonation of the quantity of TNT having the same energy release as that produced by the combustion of the flammable cloud. These assessments have been done using tables of damage similar to Table 3. Short of reviewing all the damage and conducting my own appraisal, I could have had little extra to add to the "conventional" assessments. It is clear, I think, that in most cases–with the exception of Flixborough–this review could not have been undertaken; the evidence of damage is now unavailable. Accordingly I have accepted conventional appraisals at face value. I have dwelt on the Appended theory and its implications at length, because it is novel. The only possible justification for presenting it lies in its potential correlating ability, which in consequence I have felt obliged to try to demonstrate. Correlation is not perfect, therefore predictions of the theory may be suspect. Nevertheless, as a tool it is at least as useful as others which have been applied to the problem, and in some directions its predictive potentialities are superior to any which have been developed hitherto.

The incidents which have been considered were in no way "ideal". Not only were all ranges of flammable concentration embraced, mostly the mixtures did not uniformly vary from one concentration extreme to the other. There were random variations and pockets of high and low concentration. Further, most of the mixtures were in part two phase bearing either droplets of the flammable substance, or droplets of condensed water vapour, or both. Furthermore, the flammable clouds were either uniaxial or bi-axial or complex in symmetry, but *never* spherically symmetric. Moreover, in no case has it been established that ignition occurred well within the flammable cloud; indeed *all* the reliable evidence which exists is contrary to that conception and supports the proposition (which is intuitively self-evident) that ignition always occurs at an edge.

From these comments it might be argued that experimental simulations of blast would be an impotent excercise if ideal conditions were created for test purposes. Such a view might have been correct–indeed it now appears that it may be correct–but as a scientific exercise, where of necessity the number of independent variable parameters is kept to a minimum, it has been necessary to attempt to achieve blast in an ideal situation.

Table 3 – *Peak Pressure/Damage Effects*

Pressure kPa	Damage
0.2	Occasional breaking of large glass windows already under strain
0.7	Breakage of windows, small, under strain
1.0	Typical pressure for glass failure
2.0	"Safe Distance" (probability 0.05) No serious damage beyond this value. Missile limit. Some damage to house ceilings $>$ 10% window glass broken
2.8	Limited minor structural damage
3.5-6.9	Large and small windows usually shattered; occasional damage to window frames
4.8	Minor damage to house structure
5.2	Breakage of small windows, not under strain
6.9	Partial demolition of houses, made uninhabitable
6.9-13.8	Corrugated asbestos shattered. Corrugated steel or aluminium panels, fastenings fail, followed by buckling wood panels (standard housing) fastenings fail, panels blown in
9.0	Steel frame of clad building slightly distorted
13.8	Partial collapse of walls and roofs of houses
13.8-20.7	Concrete or cinder block walls, not reinforced, shattered
15.9	Lower limit of serious structural damage
17.3	50% destruction of brickwork of house
20.7	Heavy machines (wt 1350 kg) in industrial building suffered little damage
20.7	Frameless, self-framing, steel panel building demolished. Steel frame building distorted and pulled away from foundations
20.7-27.6	Frameless, self-framing steel panel building demolished. Rupture of oil storage tanks
27.6	Cladding of light industrial buildings ruptured
34.5	Wooden utilities poles (telegraph *etc*) snapped. Tall hydraulic press (18 000 kg) in building slightly damaged
34.5-48.3	Nearly complete destruction of houses
48.3	Loaded train wagons overturned
48.3-55.2	Brick panels 0.2–0.3 m not reinforced fail by shearing or flexure
62.1	Loaded train box-cars completely demolished
69.0	Probable total destruction buildings. Heavy (3200 kg) machine tools moved and badly damaged. Very heavy (5500 kg) machine tools survived
2000	Limit of crater lip

VI
EXPERIMENTAL WORK

Broadly the concern here is with those cases, where an attempt has been made to obtain and record blast in which the oxidiser was air. Work with oxygen as the oxidiser is discounted, although not because it is necessarily irrelevant–indeed the apparent inability of moderately large-scale experiments using oxygen to generate high pressures[129] (not more than 0.08 MPa at the focus) may be considered by some to be highly significant. It is discounted as being of less direct relevance.

One factor that is immediately apparent is the lack of relevant test data. Only a handful of workers have published in the open technical literature and none of the published results support the proposition that severely damaging blasts can occur. Special reports have been published[130,131] which deal with large-scale testing of peripheral matters such as dispersion and other issues, but which are not strictly germane to the blast problem.

Some work has been done of a non-military character with vapour or gas/air mixtures ignited by detonator[132,133]. The relevance to UVCEs of these tests, as in the cases where oxygen was used, is not clear. It might be thought that as an agency for producing pressure in a gas a gaseous detonation is the most effective. Yet it has been shown[134] that pressures during spontaneous transition to detonation may exceed those found in the fully transformed situation. Whether or not detonation occurs is arguable. The present majority view, however, is that detonation *per se* does not occur in UVCEs, although the flamespeeds are undoubtedly of a transition magnitude. Nevertheless, it is not suggested that detonation cannot occur–indeed if a prolonged "pinch" is created by the geometry of the situation, it may be predicted (the Appendix) that detonation *will* occur.

One report[135] describes experiments with ethylene and air and cites the results of small-scale tests performed in a 3-m cube, whilst concentrating on the large scale experiments done in a "building" measuring 7 m by 22 m by 4.2 m to the eaves and 5.8 m to the ridge. The sides and ends of the building were covered–rather less than half the length of one end by zinc and steel sheeting and the remainder by polyethylene sheet.

Two *large-scale* experiments were performed; in the first the mean ethylene concentration at the time of ignition was 5.2% and in the other the concentration averaged 7.4% at ignition. The time delay between gas injection and ignition is not mentioned although a 15-s delay is referred to for response of the analytical equipment. Whereas the delay would have exceeded 15 s, the point is made in the report that natural ventilation (although the structure was totally closed) continually dissipated the ethylene. It is unlikely, therefore, that an extended delay occurred; nevertheless residual turbulence would have been small. Turbulence was introduced, however, by the use of seven wire mesh screens each measuring 2 m by 4 m placed within the structure. Ignition was by 20 charges of pentolite, one for each experiment.

Blast was measured by detectors located in a plane parallel to the ground and within a semicircular area, embracing the polyethylene shrouded end of the enclosure. Peak over-pressure profiles were constructed from the results. There is little doubt that a well-developed detonation occurred. The measured flame speed at the end of the structure (but within it) opposite the firing end was 1400 m s^{-1}. The measured overpressures were not in accord with expectations. The highest peak overpressures at various distances were not on the axis of the building, except for one value at 1 m from the end of the building; they were consistently recorded in the directions perpendicular to the direction of propagation; ranging from 3.9×10^6 Pa at 1 m to between 9.8×10^4 Pa and 1.5×10^5 Pa at 50 m, whereas at 50 m on the axis of the detonation the pressure had fallen to 2.9×10^4 Pa.

The peak overpressure data was plotted in the conventional manner: overpressure *vs* distance (weight of ethylene)$^{1/3}$ –and compared with TNT. Scaling the effect to one "ton" (tonne?) of *ethylene/air mixture*, the researchers concluded that at 10 m the effect was of 0.36 "tons" of TNT, at 20 m the effect was of 1 "ton" of TNT and at 50 m the effect was as if 2.7 "ton" of TNT had detonated, the distances being "from the house". Whether this was from the "centre of blast" or from the walls of the structure is not clear, but it is plainly relevant. On the basis of these figures, however, the explosion yield increased with distance as follows: 50% at 10 m, 139% at 20 m and 374% at 50 m. If what was meant in the article was one "ton" of ethylene in air, then the figures become respectively 3.5%, 9.7% and 26%.

Whether this was a real effect is naturally of some interest and requires consideration of the behaviour of the blast detectors. These were foil-type detectors and would have responded sharply to a shock wave although less well, if at all, to a gradual or even rapid pressure rise. The foil detector matrix was augmented by a piezo-electric device known as a *Tourmaline* gauge. It would appear from the paper[135] that the foil and piezo-electric detectors were in agreement. One interesting factor is that in the two piezo-electric pressure traces, obtained at 10 m and 20 m, the duration of the pressure pulse at 10 m (> 0.030 s) was two to three times as long as at 20 m–although this is contradicted in Table 6 of the report which gives the 10 m pressure duration as 0.0157 s.

The report also details damage to various constructional forms from which barriers were made and installed close to the test structure. It describes the effect upon "glass" of various types and thicknesses. Plate glass, "tempered" glass, steel-wire-armoured glass, polyester glass and acrylic glass were used, all of 5 mm thickness. Laminated glass of 6 mm thickness and polyester glass of 20 mm thickness were also tried. These glasses were set in reinforced concrete barriers containing window frames and located at 4 m, 25 m and 40 m from the test house. Only the 20 mm polyester glass was tested on 4 m; all except the 20 mm polyester glass and the plate glass were tested at 25 m, and all except the 20 mm polyester glass and the tempered glass were tested at 40 m. In all cases the window glasses were broken out of their frames, initially either as complete panes or fragments, to fly distances ranging from 3 m (40 m distant test) to more than 90 m (4 m test).

A comprehensive study[136] has been published of aspects of the behaviour of ethylene in the various forms found in industrial use. Briefly the matters considered in this research programme were rates of boil of liquid ethylene when spilled in a bunded enclosure, associated atmospheric dispersion patterns, pool burning and extinguishment of liquid ethylene, studies of the effects of fire surrounding insulated and uninsulated tanks containing liquid ethylene and ethylene gas/air explosion tests. The evaporation from a bund has been mentioned earlier, when results from the study were cited. However, with this exception, the only aspect taken further here is the explosion testing.

These tests involved quantities of ethylene in admixture with air ignited within an enclosure measuring 4 m by 4 m by 3 m high. Two explosion tests were undertaken, each of which was ignited by a 20-g charge of pentolite. In addition to many foil-type blast detectors there were three piezo-electric transducers and strain-gauge monitored "structures". The latter were of two types, both cylindrical, but respectively of large and small height to diameter ratios.

Ethylene was admitted to the test structure over a period of 60 minutes in Test 1 and had attained a mean volume of 7.2% within the bracket 6.5% to 7.9% given by eight sampling positions. Ignition caused detonation which propagated throughout the test volume. Measured pressures fell on a straight line plotted on log-log graph paper passing through 4.9×10^5 Pa at 5 m and 2.5×10^4 Pa at 30 m. Again the effect of increasing blast efficiency (compared to TNT as the standard) with distance was noted.

Test 2 failed to propagate a detonation, although the only difference was that the test chamber had been charged with ethylene over a period of 40 minutes to a mean concentration of 9.1% (the variation being from 8.3% to 10.8%). Conventional combustion occurred, however, but apparently no pressure effects were registered on the instruments.

Lind[137] has performed experiments in which conventional ignition sources have been used to ignite hemispherical volumes of hydrocarbon vapour and gas/air mixtures and ethylene/oxide/air mixtures contained in hemispherical balloons of 10 m and 20 m diameter. Ignition was at the centre of the symmetry. At the time that Lind's paper was published, ten experiments had been done, the results of which he reported. It is inferred that Tests 1 to 7 inclusive were in the 10-m diameter volume and consumed methane (5 and 7) and propane (1, 2, 3, 4 & 6), whereas Tests 8 to 10 inclusive were performed in 20-m diameter volumes and employed ethylene oxide.

Graphs presented in the description show the progress of the flames and thus their speeds. Lind found that vertical flame speeds exceeded horizontal values by 21% for methane, 33% for propane, and 60% for ethylene oxide. The arithmetic average values of maximum flame speed in each case were respectively 6.4 m s^{-1}, 9.65 m s^{-1}, and 19.1 m s^{-1}.

No pressure records are presented in the reference, although it is stated that

Photograph 1 – *This is an illustration of the sudden release of large quantities of fuel into an existing fire situation (case No 15 – California Refinery Company). Left behind on the ground, close to and within the area which originally formed the base of the rocketting tank, were most of the contents of the tank. In this case, despite the sudden release into a fire situation of an enormous quantity of readily flammable material, no explosion occurred. (Reproduced by permission of the National Fire Protection Association)*

Photographs 2 and 3 – *Damage to structures and refinery equipment close to the centre of the explosion at Pernis. For the location of the areas which are the subject of these photographs reference should be made to Figure 4. (Shell copyright)*

Photograph 3

Photograph 4

Photograph 5

Photographs 4, 5 and 6 – *These illustrate the degrees of damage at different locations from the centre of blast at Port Hudson. Reference should be made to figure 5 for their exact locations relative to the centre of the explosion. (National Transportation Safety Board, United States Government copyright)*

Photograph 6

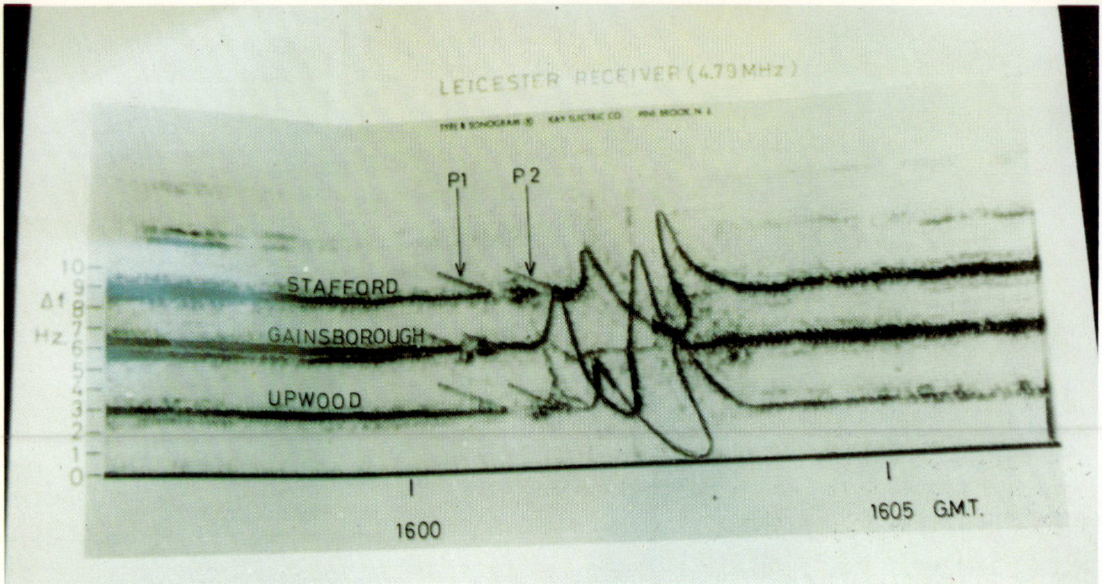

Photograph 7 – *The Leicester University ionospheric reflection records of the Flixborough explosion. The three traces appearing in this photograph correspond to the signals received at Leicester from the three radio transmitters at Stafford, Gainsborough and Upwood. The changes in position shown in the traces correspond to motion of ionised layers of the upper atmosphere. Indicated at P1 and P2 are the disturbances associated with a precursor to the main explosion. (Photograph reproduced by kind permission of Dr T.B. Jones, Leicester University, and "Nature", Macmillan Journals Ltd)*

Photograph 8

Photograph 9

Photographs 8 and 9 – *These show massive deformations of lamp-posts, which indicate the direction from which the blast wave came. (Photograph 9 reproduced by kind permission of Dr J.H. Burgoyne and Partners)*

Photographs 10 to 15 – *These show the effects of pressure on hollow vessels. Photographs 10, 11 and 15 indicate the effects of pressure on weak vessels. The subjects of photographs 10 and 15 – weak containers – ruptured in the crushed condition and thus remained almost fully deformed. In Photograph 11 the weak container did not rupture and thus its original dimensions were to some extent restored. Photographs 12, 13 and 14 show the effects of the blast at Flixborough on the skirts of strong vessels. In these cases some degree of pressure equilibration tended to occur during the blast because of gas flow through the apertures in the skirts of these vessels. It is this aspect which has enabled an estimate to be made of the minimum rate of pressure rise in the atmosphere necessary to cause this crushing damage. (Photographs 10 and 11 reproduced by kind permission of Dr J.H. Burgoyne and Partners)*

Photograph 10

Photograph 11

Photograph 12

Photograph 13

Photograph 14 ▷

Photograph 15

Photograph 16

Photographs 16 and 17 – *These show a cast-iron manhole cover located intermediately between the reactor shown in photograph 12 and the road tanker shown in Photograph 15. This manhole cover was broken by pressure alone*

Photograph 17

Photographs 18 and 19 – *These illustrate the damage in reactor No 6 at Flixborough, which appears diagramatically in Figure 9. In photograph 18 may be seen the bend on the stirrer shaft and in Photograph 19 what had been the baffle plate, turned inside-out by its impact with the stirrer shaft. (Reproduced by kind permission of Dr J.H. Burgoyne and Partners)*

Photograph 18 ▷

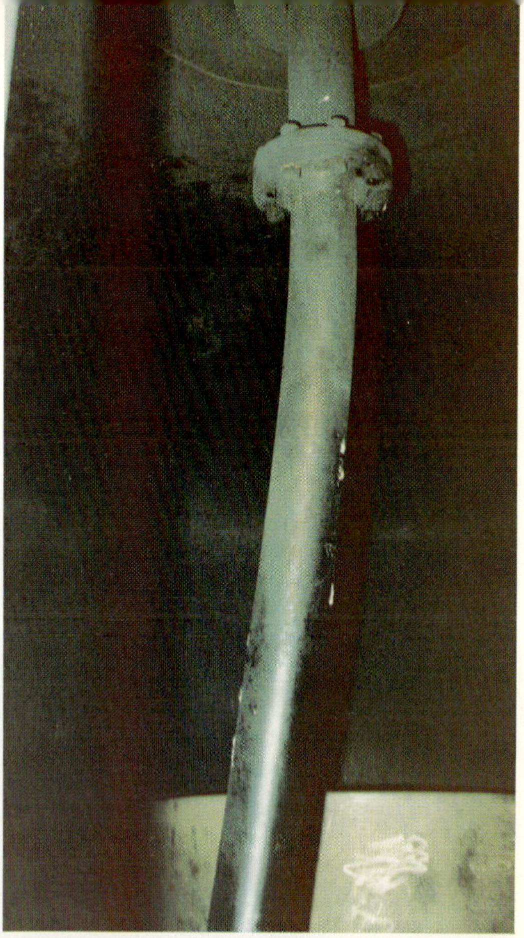

Photograph 19

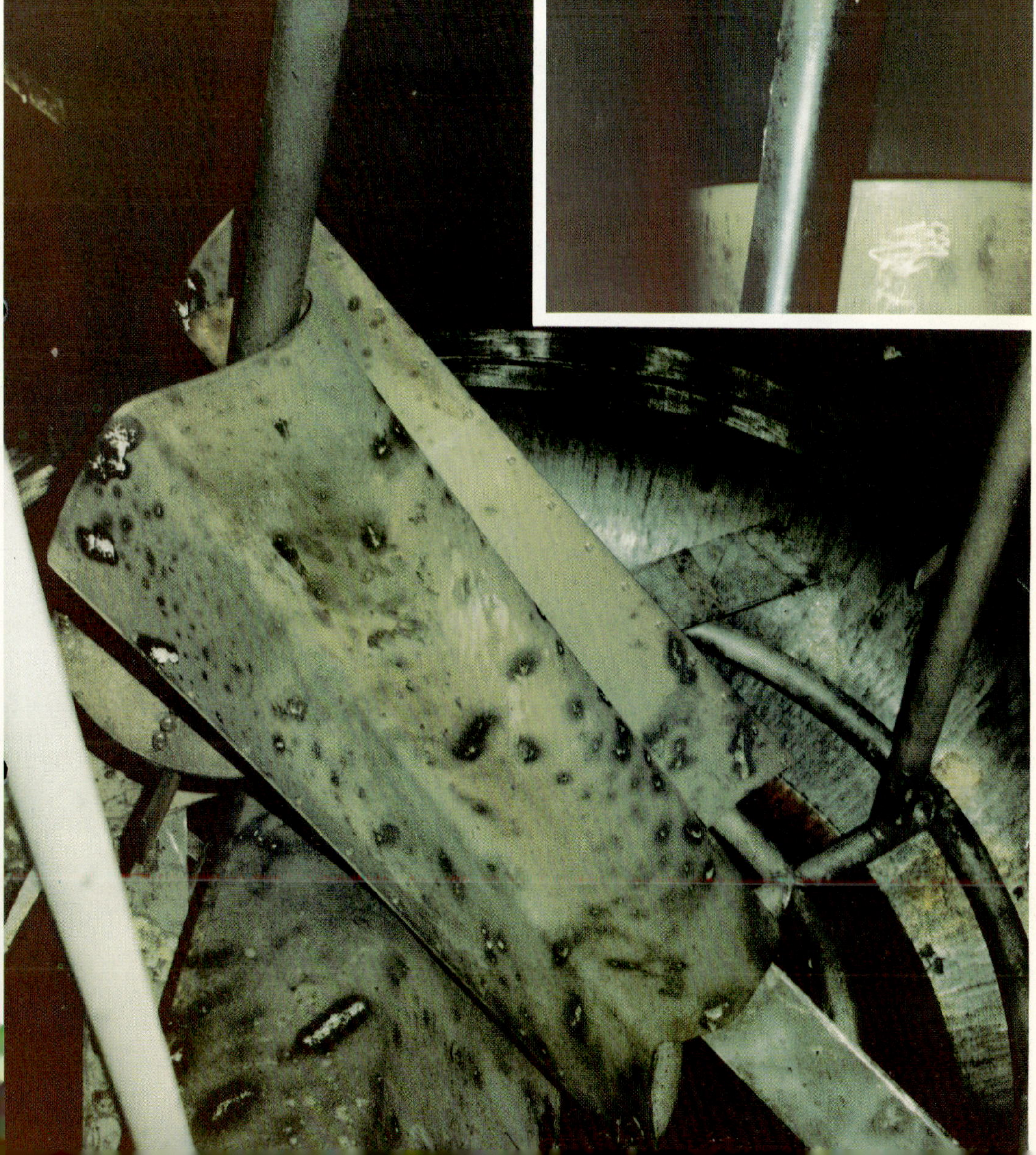

Photographs 20 to 25 – *Views of lamp-posts on the Flixborough site; the even numbers correspond to views in the plane of distortion and these indicate the direction of the blast wave when distortion occurred, and the odd-number photographs show views of the lamp-posts at right angles to the plane of distortion, from which may be gained some impression of the energy dissipated by the blast wave at that location. (Reproduced by kind permission of Dr J.H. Burgoyne and Partners)*

Photograph 20

Photograph 21

Photograph 22

Photograph 24

Photograph 23

Photograph 25

Photograph 26 – *A view of a lamp-post close to the centre of blast, indicating the non-uniform character of the blast wave impinging from two or more different directions. (Reproduced by kind permission of Dr J.H. Burgoyne and Partners)*

Photograph 27 – *A lamp-post as close to the centre of the blasts as that shown in Photograph 26, which had experienced forces, the net sum of which had led to minimal distortion. This lamp-post lay approximately on the line between those shown in Photographs 8 and 9 and may be seen in the background of Photograph 8. (Reproduced by kind permission of Dr J.H. Burgoyne and Partners)*

Photographs 28 to 31 – *Damage to plant items at the positions indicated in Figure 10. Photograph 28 shows one of the possible points of emission of propylene. Photographs 29, 30 and 31 show damage which was used diagnostically to evaluate the yield of this explosion in terms of TNT, respectively failure of reinforced concrete panels, collapse of the dished end of a vessel and the lifting of a reinforced concrete cantilevered slab at the edge of a plant support structure. (Photographs reproduced by kind permission of the Ministerie van Sociale Zaken, Netherlands)*

Photograph 26

Photograph 27

Photograph 28

Photograph 29

Photograph 30

Photograph 31

pressures were recorded in the 10-m (radius) test and the results are currently being assessed. No pressures were recorded in the 10-m diameter tests. This was due to the inability of the piezo-electric system to detect less than 0.1 bar (10 000 Pa). The author makes the point that such low flame speeds would not have been expected to generate measurable pressures on the basis of the theories of Kuhl, Kamel and Oppenheim[128]. Indeed this is the case also if the appended theory is applied; the maximum pressures anticipated would have been 12 Pa for methane in the 10-m diameter balloon, 51 Pa for propane in the 10-m diameter balloon and 360 Pa for ethylene oxide in the 20-m diameter balloon.

It is significant in the context of the appended theory that pressures *were* observed with ethylene oxide—although not quantified—when the article makes it plain that these must have exceeded 10 000 Pa, the point being, if this is right, that the appended theory must substantially underestimate pressure. No doubt the matter will be clarified when these test results are published in full.

Three studies were reported in Volume 10 (1976) of *Loss Prevention* which dealt with the effects of water sprays on UVCEs. As with the other tests which have been reported, all the "unconfined" mixtures were initially contained within film-shrouded or balloon-type enclosures. One report[138] considers the effects of water sprays on dispersed ethylene and vinyl chloride in an enclosure measuring 3 m by 3 m by 4.6 m high. No pressures were measured, although events were recorded cinematographically. The authors comment that although flame speeds were quite slow during the ignition phase, they were universally augmented by the application of water. No subjective comment was made which might have indicated the development of pressure. They presented results indicating flame speeds as high as 36 m s^{-1} for ethylene.

Another article[139] reviews the results of tests devised to study (1) the detonation requirements of ethylene and hydrogen gas/air mixtures with and without water spray, and (2) the "explosive" yields of large volumes of flammable gases. It was found that, although the air pumping effect of a water spray might be useful in rendering explosive mixtures non-explosible, the mere presence of finely divided water as a medium within the flammability limits will not quench combustion.

The "yield" experiments all involved 7% ethylene/air mixtures in polypropylene bags 3 m, 6 m and 8.15 m in diameter. Pressures were measured and are quoted for one point in each of the seven tests at distances from the centres of the bags of 4.27 m (3-m bag), 7.32 m (3-m bag), 7.62 m (3-m, 6-m and 6-m, and 8.15-m bags), and 15.24 m (6-m bag). Even the smallest bags had tearing pressures of 1 inch water gauge or less, say 250 Pa.

Unfortunately the ignition source (or sources) for these tests are not specified in the reference, although it is stated that electrical sparks and explosive sources were used in the overall test programme. In five of the seven tests the conclusion was that deflagration only had occurred, but that in the other two, detonations had resulted. In the absence of specific information on the mode of ignition it is impossible to assess these results. It was calculated, however, that the yield related to TNT was in the range 0.04% to 0.87% for deflagration, was 5.94% for deflagration with partial confinement and reached 54% for detonation. No flame speeds were reported so a theoretical assessment of the results has not been possible.

Vincent and Howard[140,141] detail the effects of water sprays on hydrocarbon mist explosions. Typically they considered mists of relatively involatile materials such as heat transfer liquids. The large-scale test enclosure measured 6.7 m by 4.9 m by 4.6 m high. Again measurement of pressure was not attempted, although the authors made the following comments. "All explosions were deflagrations and in no case was a detonation experienced. Thus, the explosions were not accompanied by loud noises or shock waves. Because of the very large vent area (on walls and ceilings of the structure) no appreciable pressure is believed to have developed."

The only experiments of which I am aware in which dispersion and ignition have been attempted in truly unconfined situations are those presented in the series of papers emanating from BASF, Ludwigshafen, West Germany. One[142] deals with the mechanics of failure rather than combustion matters, although these are mentioned, and another[143] considers at length the explosion phenomenon.

In both papers the method of creating the flammable cloud is described; superheated flammable liquids, respectively propane and propylene, are released into the surrounding atmosphere by the initiated but otherwise spontaneous rupture of their containers. In the first experiment the containers were small cylinders having length-to-diameter ratios of 3.5:1 and volumes of 1.76×10^{-4} m^3 and 5.94×10^{-4} m^3. Their wall thickness was respectively 0.3 mm and 0.5 mm and for a reproducible *point* or *line* of failure, they were scored from end to end. To achieve failure at a given point of time, however, the test cylinders were instantaneously shock loaded by gas pressure. This was done when the test cylinder pressures had reached 8 to 9 MPa.

The later tests were very similar in the way the vapour cloud was generated, except that mechanical or explosive rupture initiation was adopted when the pressure of the contents was about 6 MPa. In these tests the real differences lay in their scale compared to the earlier tests, although the small-scale tests were included. The test volumes this time had length-to-diameter ratios of 4.5:1 and were 2.26×10^{-4} m^3, 7.63×10^{-4} m^3, 3.53×10^{-3} m^3, 1.19×10^{-2} m^3, 2.83×10^{-2} m^3 and 1.21 m^3 in volume.

In the combustion experiment, ignition was achieved by the discharge of a capacitor across a number (usually 12) of thin wires such that they exploded. These wires were arranged in a straight line along a radius to the test container. Pressures were measured by pressure transducers and visible phenomena were recorded on moving film. The two sources of pressure effect had to be differentiated: rupture and combustion. In these tests the former exceeded the latter quite significantly in all cases. Nevertheless there was no interference between them as the rupture shock was beyond the test zone before ignition was attempted.

Measured pressures "at the edge of the vapour cloud" ranged from 0.5 to 1.5 kPa for the smallest vessel to 4 to 7 kPa for the largest. The corresponding observed flame speeds were from 10 to 25 m s^{-1} for the smallest to over 48 m s^{-1} for the largest. It is of interest to note the relatively low value of pressure found in the largest test. The authors extrapolate from this to 1.2×10^4 Pa for 50 tonnes of propylene having a flame velocity of 100 m s^{-1}.

The value of pressure found by the authors was well below that predicted from conventional theory for a flame speed of 48 m s^{-1}. Kuhl, Kamel and Oppenheim indicate (Figure 11 of their paper) that the anticipated overpressure arising from such a flame speed would be 7.1×10^4 Pa (71 kPa). Moreover at 100 m s^{-1} this would be increased to approximately 2.5×10^5 Pa (by estimation from Figure 11[128]).

It is recognised that this theory cannot be rigorously applied to any situation involving flame acceleration, nor one in which the character of the combustion process is variable from point-to-point in respect of mixture concentration and turbulence. Nevertheless it was the only predictive tool available until recently. However, analysing the largest test result by the theory given in the Appendix suggests that the maximum pressure which would have been expected in the test at a flame speed of 48 m s^{-1} would have been 5.42×10^3 Pa, which is in fair agreement with what was observed. However, applying the same theory to 50 tonnes and 100 m s^{-1} flame speed gives a maximum pressure of 4.90×10^4 Pa, or four times that extrapolated by the authors[143].

Compared to the pressures which have undoubtedly occurred in UVCEs, these forecasts are modest. Some enhancement of flame speed beyond 100 m s^{-1} might be expected. Nevertheless, as was indicated earlier, the situation is rather less conducive to the development of an extended cloud and much of the discharge probably remains above the upper limit of flammability. If ignition is soon after the rupture a fire-ball would probably result. The comments made in the Appendix apply here; as the cloud was symmetrical, whenever the source of ignition was applied the development of the "pinch" effect with its concomitant flame acceleration could not arise.

The test situation was a little unreal in the context of most UVCEs. The BASF Ludwigshafen explosion, Case No 14, seems to be the only one where instantaneous and *complete* disruption of the container may have happened. The evidence in all other, apparently similar, cases suggests that sufficient of the tank remained even if only momentarily (case 38) to direct the emission linearly and thus achieve a favourable geometry for blast. Indeed such complete and instantaneous failure may not have occurred at Ludwigshafen. Insufficient evidence is available to discriminate. Perhaps the tests were devised to eliminate symmetry as a variable. If they were so devised, it is probable that the low pressures obtained were a consequence of that decision.

Over the years operators of high-pressure ethylene polymerisation processes have suffered aerial explosions varying in severity and frequency, depending upon the mode of operation. These have been known to break windows at some distance, although they are less of a threat than a nuisance. Simultaneously over the years attempts have been made to prevent these explosions either by suppressing ignition or suppressing combustion. It has never been accepted in the industry that emergency dumping from such reactors can be abandoned nor even that safe discharge to dumping vessels would suffice. "Safe" in this context means that no increase in hazard to the reactor is experienced.

Three factors characterise polyethylene reactor bangs. One is the extremely rapid rate of mixing of ethylene with air. Another is the form of spontaneous ignition which occurs, which has the characteristic of simultaneous ignition throughout the volume rather than normal or even detonative flame movement. The third is the relatively small quantity of ethylene involved–some tens of kilograms only. In the course of the study of the problem a number of approaches have been adopted.

Marzais of the Société Ethylène Plastique, Paris[144], considered water injection and ignition suppression. Here the aim was to lower the existing gas temperature below its spontaneous ignition value (and, as a bonus, add a diluent) when mixed with air. Marzais' studies of the physics of the process were related merely to the mechanics of mixing and subsequent heat and mass transfer. No combustion studies were undertaken.

Those who have approached the problem on the basis of combustion suppression, however, have studied ignition and combustion behaviour[145]. For convenience, as a discharge of decomposing ethylene will not reliably ignite spontaneously, and as contrived ignition does not produce the same effect, this was approximated by the rapid release of high-temperature carbon monoxide and hydrogen from a high-pressure bomb. The high temperature and pressure were achieved by first igniting a mixture of oxygen and methane, initially also at high pressure, and discharging these rapidly to atmosphere by means of bursting disc. The issuing gases were at a temperature high enough to spontaneously ignite on contact with air. It is understood that the quantity of flammable gas involved was only a few kilograms and that the pressures measured at the centre of the burning cloud were about 2.5 MPa.

Such high pressures might be predicted, if 5 kg of methane had originally been involved, if the *effective* flame speed had been more than 1540 m s^{-1}, or, alternatively, if the time of emission together with the ignition induction period had been less than 0.00167 s. This is very rapid combustion and the above figures indicate that the "effective" flame speed was very high—comparable with detonation velocities in carbon monoxide-hydrogen/air mixtures.

The results of the experimental work are equivocal. There would seem to be no doubt—based strictly on experimental evidence—that plant designers and operators could for the most part claim that there was very little, if any, evidence that severely damaging UVCEs could or would happen. The most obvious limitation of the experimental work has been in the scale of testing. The largest test to date has involved only 452 kg of material[143]. Nevertheless mass does not appear to be the sole criterion; another criterion is flame speed[128]. The intrinsic reactivity of the flammable concerned may also be thought relevant, and the most obvious combustion manifestation of this is burning velocity. Yet all types of materials, it would appear, can generate UVCEs—even those with low intrinsic reactivity. Contrarily those with high reactivity, *eg* hydrogen—when pure, are reluctant to explode with force.

It seems that hitherto burning velocity and flame speed have been dissociated from the aerodynamic context with the result that potentially massive flame acceleration effects have been overlooked. Whereas turbulence induced by flow undoubtedly contributes to events, it does not appear to be wholly capable of explaining them. I would recommend that—in any future programme of testing—"edge" ignition effects should be studied in systems with uniaxial or biaxial symmetry, but not those in which three-dimensional symmentry exists. I would suggest also that the effects of molecular weight of the flammable and its temperature upon release (or after mixing with air), and physical state, be studied. Such tests with sensitive instrumentation could undoubtedly produce worthwhile positive (or negative) results even on a modest scale. An examination of situations in which substantial proportions of the volume of the flammable cloud (or the surrounding areas) comprise non-compressible matter might also be worthwhile. That is the effects of dense clouds of droplets at one end of the scale and dense plant accumulations at the other. Well established theory and sound experimentation[146, 147] have demonstrated that both are capable of affecting the speed of sound in a way which could only intensify pressure effects; any reduction in the local velocity of sound would produce similar results to an increase in flame speed.

I would advocate also detailed attention to the wave form generated by experimental UVCEs and the effects of, for example, point of ignition, or terrain, or wind, *etc*, on any polarisation of pressure effects there might be. Particularly, attention to duration of pressure is important as control rooms and other buildings are and will be under study for construction. For rational design, such structures require detailed impulse data.

VII
DIAGNOSTICS

It is well known that the effects produced upon an object by a force are dependent upon its response characteristics. The response characteristics involved here are those which lie outside the elastic, or self-restorative, ranges of behaviour. These may be simple translation or, more usually, distortion followed by rupture and translation. Now it will be apparent that all UVCEs are characterised by pressure and from this that many such events have been quantified by assessments of pressure. These assessments have not usually taken the form of the analysis presented in the Appendix; they have been based upon either explosive damage data, such as Table 3, or straightforwardly calculable loadings which have resulted in, say, tensile failure.

Both techniques are relatively unsophisticated. Patently neither is appropriate in the pressure-waveform regime associated with the other. And yet hitherto they have tended to be used interchangeably. Table 3 details damage, most of which allows a considerable degree of flexibility in selecting the appropriate pressure level. This is because most of the damage referred to is building damage. Buildings (and those structures in the Table which are not buildings) are complex structures having both intrinsic differences in strength (due to age, type of materials and means of construction) even when closely similar in appearance, but generally they are all different and difficult to strictly compare on the basis of damage.

Table 3 contains data compiled from many different sources, including the United States Army, Brasie and Simpson[107], United States Bureau of Mines and Glasstone[148], and is appropriate only for shock-wave damage. What is missing from this catalogue is any definition of terms such as "large", "small", "under strain", "minor", *etc.* As all the effects quoted are pressure effects, plainly the areas of damaged items are relevant. Equally relevant are their thicknesses. Deformation and failure under "static" loading conditions is not usually divorced in this way from such relevant criteria.

Following serious explosions in buildings, the Building Research Station instituted and conducted a programme of experiments to quantify their effects. Reports on these have been issued[149, 150]. Whereas the purpose of the experiments was to examine the behaviour of different internal pressure loadings on, for example, walls of traditional construction but of different thicknesses and spans, no real departures from the observed behaviour would be anticipated from similar (non-shock) pressure loadings applied externally.

No less than six categories of pressure are associated with window breakage in Table 3, ranging from 0.2 kPa to 5.2 kPa. Clearly a great many styles of window are included, but it seems unlikely that different *types* of window are considered; that is, where differences relate to steel-wire-armoured, toughened or otherwise modified glass, rather than merely to size and thickness. The lack of precision which this range of pressures allows could result in analysts differing by as much as one order of magnitude in the amount of explosive which created the shock wave. This imprecision is most unfortunate, as broken glass is potentially by far the most useful diagnostic tool. The damage is conspicuous and, because it occurs at lower pressure levels than cause structural damage, it is usually unequivocal. Further, because it is low-pressure-level damage, it is likely to occur in an annular region of greater area and thus give statistically significant data—not only on explosion severity but on blast wave polarisation, *etc.*

In all the assessments of UVCE explosion yield presented in Table 1, glass damage will have figured largely. And yet it is a material which has been subjected to few tests. Glass panels have been used in explosion relief experiments[151], but what was observed and reported was chiefly the effect on the course of the explosion and not the effect on the glass—saving that in every test the glass was broken. Other than the tests run by RVO-TNO[152] and Pittsburgh Plate Glass Co[153], few quantitative data are available[154, 155]. The most recent appraisal has been by the British Gas Corporation[156]. In this work 3-mm, 4-mm and 5-mm plate glass measuring 1 m square, 1 m by 0.5 m and 0.48 m square was tested, as well as georgian-wired glass (6.5 mm thick). The glass window specimens were rough-cast or polished and treated or untreated ("treated" glass in this context was with the application of 3M (UK) Ltd shatter-resistant film).

This work has shown that the breaking pressure of a glass panel (for squat, rectangular areas) is inversely proportional to its area and proportional to its thickness for all the types of glass tested. The authors consider their results to be as appropriate to external pressurisation as they are for internal pressurisation, provided the incident pressure wave form is similar. It is true that the work would not be strictly applicable either to steep-fronted pressure waves nor, necessarily, to quasi-static loading. The range of breaking pressure found in these experiments was from 2 kPa to 23 kPa.

The specific influence of shock-waves on glass (and other materials) has been studied by Breitwiser and Grimmingen[157]. They detonated polyethylene balloons of 1.2 m diameter filled with hydrogen and oxygen at distances of 1.5 m and 3 m from test specimens mounted in a framework about 1 m square. The calculated impulses attained in this work were roughly an order of magnitude less than those in the British Gas work. Nevertheless glass 7 mm thick was broken, which possibly reflects the relatively high maximum overpressure which was created, namely 91.2 kPa. The mode of breaking was also different and *diagnostically significant* in that cracks were largely circumferential rather than radial. The British Gas Corporation report[156] also alludes to this matter.

Of all the diagnostic techniques, the analysis of motion of a simple unrestrained object is probably the most rewarding in terms of differentiating between the various possible pressure-wave forms. This analysis can probably best be applied to an animal or human body as it is relatively unusual to encounter a totally unrestrained but otherwise freely mobile, inanimate body. The Appendix presents the relevant theory for the two extremes–explosive shock-wave and the "sinusoidal" form of blast wave. The former is supported by data obtained from massive TNT detonations[158] in which the pressure wave duration at an overpressure of 3.66×10^4 Pa (5.3 psi) was 0.96 s. In this case a 79-kg (169-lb) man was still accelerating at 0.5 s and had involuntarily moved through 2.5 m (8 ft) whilst achieving a maximum velocity of 6.5 m s^{-1}. By way of example, relevant to Flixborough, the detonation of 18 tonne of TNT should at a distance of 200 m have been capable of accelerating a man to 0.5 m s^{-1} in 0.04 m.

As is clear from the Appendix, the main translational agency in the case of explosives is wind. The shock front is past in too short a time to have a significant effect on all but the smallest particles. For the blast wave model presented herein, a different situation exists. No distinction has been made in the model between "static" and dynamic pressure effects–the two have been lumped together, although both effects would no doubt be distinguishable. For a pressure peak in the travelling wave of 10^5 Pa with positive and negative pressure gradients of equal duration and totalling 0.6 s, then the displacement of a 70 kg man would be about 0.2 m only, but as he would have been effectively pushed in the back for 0.3 s by a force of some 150 N, it is highly probable that he would tumble, especially if he was running.

Workers have approached explosive blast-wave effects in different ways. Bowen *et al*[158] have taken the view that wind effects alone are responsible for translational movement. Baker *et al*[159], however, express effects in "force" terms where the terminology clearly implies pressure force. Of course, wind exerts its energy by producing a pressure effect, the dynamic pressure, therefore the two approaches would, on this basis, appear to be identical. There is, however, an element of pressure in the explosive blast wave which is not associated with wind–the shock overpressure. On the face of it, one approach would seem to include that pressure whereas the other would appear not to.

Bowen *et al* have experimentally verified their theoretical work, modifying acceleration coefficients to fit only where tumbling of elongated objects was involved. No equivalent experimental correlation has been produced for the action of a pressure force alone.

Diagnostically, the movement of a free object is of use only if its position prior to the incident is known with certainty. It is unlikely that this will ever be the case with any person due to the general lack of objectivity which prevails during such incidents. Accuracy is clearly necessary in view of the small displacements involved (for bodies with low acceleration coefficients), 0.08 m and 0.2 m for similar distances from "equivalent" blast foci.

In the Appendix are presented two techniques for appraising blast effects. Lamppost deformation has interesting possibilities, which go beyond mere blast direction indication. The energy expended on the lamppost may be calculated. This energy (the energy in *excess* of that which will *just* not cause permanent deformation) is provided by that part of the blast wave impinging on the lamppost. Thus the total energy of the blast wave, at that particular radius from the focus of blast, may be assessed by proper proportioning. The second technique involves appraisal of the damage to relatively strong containers. This is particularly useful because it yields minimum figures for the necessary rate of pressure rise and the final crushing pressure. It is the only technique which enables rate of pressure rise to be assessed. For this reason it is potentially the most important diagnostic tool available at present.

VIII
MISCELLANEOUS ISSUES

In this section I propose to deal briefly with a few matters of relevance, some important, others less so: effects of wind and weather generally, effects of terrain, "anomalous" blast effects and mortality. Of these the most important is clearly mortality.

Mortality

No case is known in which any person, or indeed animal, has been killed unequivocally as a result of direct (primary) UVCE blast effects. All those who have died have suffered indirect (secondary and/or tertiary) effects. The latter includes missiles, whole body displacement and direct and indirect thermal radiation or fire effects.

Indeed fatal injuries would not be expected from UVCE blast waves alone, in view of the tolerance of animals to "slowly" rising pressures, *ie* up to 10^7 Pa s^{-1} (the Appendix), and the evidence (the Appendix) that rates of pressure rise are unlikely to exceed that figure and that high pressure would not last more than a few tenths of a second. Nevertheless, the issue of surviving a UVCE blast whilst actually being within it has hitherto been academic. Even if buildings strong enough to withstand UVCE blast are erected, no one within them is going to be exposed to the external atmosphere during a blast and survive because of the undoubtedly lethal effect of wind, high temperature and toxic products of combustion. The advent of strong buildings therefore will not alone render survival more likely.

How people die is relevant only to preventative measures. Their deaths focus attention on those aspects of their environment which perform unsatisfactorily in UVCEs. Invariably this unsatisfactory performance relates to buildings or structures. In due course it is to be hoped that plants will be constructed which will provide adequate protection against injury or death in the event of UVCEs.

What is useful now is an indication of the real risk of death associated with the current generation of chemical plant. The problem is complicated by two basic factors: the wide range of types of chemical plants currently operating and the extremes of environment in which they are located. The first matter bears on the probability of an incident as well as the possible mortality. The second relates largely to mortality alone.

Old plant tends to be smaller in terms of size of unit operation undertaken. Because it is older, however, it will generally be more prone to failure. Also, older plant tends to be located within the community upon which it depends for labour. The most modern plant, on the other hand, tends to be less closely integrated although still proximate to the community. The reasons for this are commercial and environmental rather than directly related to explosion hazard—although they may have included, for example, an appraisal of toxic risks. Hitherto, in-plant explosions have not been considered to present a significant threat to the surrounding community. Damage which has occurred as a result of most in-plant incidents has been due to projectiles. Flixborough was the first incident in this country to demonstrate that blast was a potent source of danger to the community, albeit that not one member of the general public was killed in that event. The significance of this will be considered shortly.

Marshall[56] has recently produced a Mortality Index for a variety of "incidents" (including war) where explosive effects have caused death. His correlation is based upon 162 accidental detonations of conventional explosive ranging in scale from 5 kg to 2000 tonnes in which 5809 persons were killed. He grouped the data into twelve bands of explosive mass (separately grouping the three lowest bands and the highest bands together) and plotted the resultant nine points on a graph of fatalities per tonne versus tonnes of explosive involved. Marshall has called his "fatalities per tonne" the Mortality Index and found the following correlation:

$$\text{Mortality Index} = 4 \times (\text{tonnes of explosive})^{-1/2}$$

Deviation from this "best-fit" equation is more pronounced for the large conventional explosive incidents used in the correlation, but the deviation would not appear to be significant as judged by these incidents alone. Significance may be attached to the deviation, however, when points are plotted for the following non-conventional "explosive" detonations: Nova Scotia (1917), Oppau (1921) and Hiroshima and Nagasaki (atomic bombs). These four incidents taken together with the two largest conventional explosive detonations could indicate a trend towards a correlation parallel to the above, in which the coefficient is about 160 instead of 4. The relevance of this is that the larger explosions effectively integrate natural population movements and thus capture victims whose movements would otherwise render them safe from small explosions. It is probable also that all the largest incidents affected residential and/or industrial areas of towns (certainly most of them did) and thus affected concentrations of population, rather than rural or depopulated areas.

In predicting the effect of a UVCE, which coefficient should be used? Marshall has plotted Flixborough and the means of 16 incidents (including Flixborough) on his graph (cases 7, 9, 36, 39, 54, 55, 64, 66, 71, 76, 79, 81, 82, 92, 99 of this report and one non-ignition). Both Flixborough and the mean lie close to the curve defined by the coefficient 4. Nevertheless, had the Nypro plant been in an urban area, fatal casualties would undoubtedly have been higher. Moreover, there can be no doubt whatever that had this accident occurred during office hours on a weekday, fatalities would have been much greater. Of course, there can be no certainty about just how much greater; counting the two factors, however, 300 fatalities would appear not to be an unreasonable estimate for the worst case.

It is perhaps fortunate that the "worst case" does not often arise—indeed it is the exception—nevertheless for planning, protection and prevention purposes it must be assumed. It is certainly foreseeable without undue imagination. Marshall has plotted the sizes of the UVCE incidents on his graph without correction to TNT equivalents. That the correlation should work on this basis is interesting although the data scatter scarcely warrants the use of equivalents. In any case direct plotting tacitly assumes in each case (for hydrocarbons) an efficiency of 10%.

Flixborough plainly illustrates the potential of these explosions. Furthermore, it indicates that for planning and other purposes the Mortality Index curve should start to deviate from the lower correlation towards the upper for incidents in the range of from one to ten tonnes. This conclusion is reinforced by case 14, which generates a point on the mortality index graph well above the correlation produced by Marshall.

In fact, this is clearly a valid conclusion in respect of the BASF incident, because the Marshall correlation, if strictly applied, would indicate the involvement of 2730 tonnes of exploding material, whereas it is known that only one rail tank car was involved.

The mortality index correlation is different from that to be expected, were overpressure alone the factor responsible. However, since most fatalities are due to indirect effects and since it would appear that most structures, where explosive blasts are concerned, are sensitive to *impulse* rather than overpressure only, then working through the theory in the way Marshall has done indicates the following correlation, based upon impulse:

$$\text{Mortality Index} = \text{constant} \times (\text{mass of explosive})^{1/3}$$

This expression, it will be appreciated, could provide a justifiable mathematical relationship for the "transition" between the existing Marshall correlation (cited earlier) and the higher mortality index appropriate for larger explosive charges. This correlates with large explosive and atomic incidents if the constant referred to has the value 5×10^{-2}. The region of transition extends, on the data available, from about 300 tonnes to 30 000 tonnes. (Applying one's mind to the BASF and Flixborough situations, however, these three figures could well be 1.0, 3 tonnes and 300 tonnes respectively.)

The danger in these arguments lies in placing too much faith in the data. Insufficient data exists to put the deductions on a statistical basis and the nature of the data itself tends not to be coherent; opposite ends of the spectrum of scale are not merely different in magnitude—they are different in kind.

The basic problem is how to be *realistic*. The Marshall correlation, extended as may be appropriate to very-large-scale explosions, represents what is apparently the *real* risk. It will be clear, however, that this is to be distinguished from the maximum potential risk. How the gap is bridged is a matter of technical argument, Public Policy for the Government and capital expenditure coupled with Insurance by Industry.

Anomalous Effects

These anomalies are those observed with explosions where, at least, an identifiable centre to blast exists.

Precursors have been mentioned. These consist of information registered by recording equipment of the existence of an approaching shock wave before it actually arrives at the responding instrument. Baker[41] mentions them and explains them as either preceding aerial shock waves caused by supersonic fragments (such as shell casings) or ground waves propagating in advance of the aerial shock.

The only UVCE having precursors was Flixborough[123] (Photograph 7). As there were no supersonic fragments and as ground waves cannot explain the recorded effects (the precursor becoming less far ahead of the main wave the further from the centre of blast), these precursors were attributed by some to a real event preceding the main blast. It is true that the wave form of the precursor was markedly different in shape from the main wave and remains a puzzle.

Anomalous winds have been described; indeed they have been measured by some workers[160] whose interest has been in the translational effects of explosive detonations. These are not uniform in any sense; they neither uniformly varied in strength or direction, nor were they observed over the whole blast-affected area. Such effects in a simple geometric system with explosives indicates that similar effects must be anticipated with UVCEs. Fletcher and Bowen were unable to attribute any rational cause to the anomalous winds. A degree of rotation might be anticipated due to Coriolis effects. Undoubtedly the Coriolis force would have arisen but its action should have been uniform rather than random.

Meteorological Influences

These affect the course of a UVCE at all stages from emission through to remote pressure effects. The primary influences are wind velocity and temperature; humidity, rainfall, atmospheric pressure, *etc* are secondary. Basically the concentration at any point in a vapour cloud, whether "instantaneously" created or of an equilibrium size, is inversely related to wind speed. The relationship is a simple inverse for continuous sources of flammable; it is a power of the inverse for an instantaneous release. The corresponding volume of flammable concentrations is, at least, inversely proportional to the cube of the wind speed. Wind is therefore an important, albeit unreliable, mitigating influence in UVCEs. Another mitigating influence in cases of "instantaneous" emissions is ignition delay, which is related to explosible volume in the same way as is wind speed.

Depending upon the local prevailing wind, areas of high and low exposure risk may be defined in any plant. It must not be assumed, however, that vapours released from plant will all flow downwind. This very much depends upon the force of emission and quantity emitted. But irrespective of wind direction and force (within moderation), density gradients may arise which could cause vapours to flow against the wind even if they had no upwind velocity component at source. Another profound influence in this context is, of course, terrain.

The influence of meteorological conditions on the *effects* of a UVCE has been considered. This was a refraction study on the blast wave in an attempt to explain remote and apparently anomalous behaviour. Burgess and Zabetakis[91] adopted the techniques already developed to explain similar effects from conventional explosives, *eg* Reed *et al*[161]. Broadly the anomalous behaviour referred to was the breaking of window panes in regions beyond the ideal radial limit of pressure values high enough to cause that breakage. As it was neither convincing nor possible to explain the damage in terms of weak or large glass panes, refraction of sound waves and their consequent focussing was hypothesised. As the atmosphere does not behave as a lens, the best focus which it is possible to suggest is the caustic. This is, in fact, all that is necessary to explain the damage. Better focussing could, of course, lead to much more severe damage which has not been observed.

As with light, focussing is achieved with sound by media which alter its velocity. The two principal factors are temperature and wind velocity. In a quiescent atmosphere displaying the "normal" (adiabatic lapse rate) vertical temperature distribution sound is diverted away from the ground surface. Rarely is the atmosphere quiescent and often the "normal" temperature gradient is replaced by neutral or inversion conditions. In an inversion, temperature gradients are reversed within a few hundred metres above the ground. This, coupled with, for example, winds of different speeds or directions or both at different elevations can lead to sound being bent back towards the ground surface and even being focussed there. Reflection from the ground at the focus (or caustic) cusp has been suggested[91] as an explanation for successive anomalous damage foci.

The work of Reed *et al* was coupled with a survey in depth of glass structures at risk. Their appraisal, taking into account—as it must—"damage" not actually attributable to the explosion, coupled with data on the strength of glass provided by manufacturers, enabled them to correlate "damage intensity" (D) in broken panes per thousand with pane area (A) and overpressure (ΔP) for "single strength" glass as follows:

$$D = 3.71 \times 10^{-3}.A^{1.22}\,\Delta P^{2.78}$$

The correlation is purely empirical and is true for the overpressure values 0.25 kPa and 0.4 kPa.

Quite what "single strength" means is not explained. It is clearly a term which was used by the manufacturing company, Pittsburgh Plate Glass (PPG). By using Figure 12 of Reed, Pape, Minor and De Hart's paper[161], however, and Harris, Marshall and Moppett[156] it would appear to mean glass of thickness 1.3 mm, but this would seem to be far too thin for practical purposes. Nevertheless it would explain the very low breaking pressures compared to those cited earlier—Section VI. However, it is important to emphasise that the pressures used in deriving the correlation were calculated on the basis of caustic theory expounded in that reference[161].

It is clear that refraction of sound waves does help to explain otherwise anomalous blast effects both in respect of explosive detonations and UVCEs. Refraction theory, as yet, does not provide the whole answer, however. Terrain and coupling ground reflection effects may assist in developing the whole picture. It is unlikely that such factors will be of more use than to explain damage which has already occurred, predictions of such damage could be too erratic to be of much help.

Terrain is *a priori* an important influence in UVCEs, both for the pre-ignition phase of the event and in subsequent blast. Two cases already considered in detail—Port Hudson and Flixborough—were in, or adjacent to, hilly areas. Both gave apparently anomalous or polarised blast effects. One report[91] considers terrain only in respect of disposition of the vapour cloud before ignition. In that case propane issued from the buried pipe close to the head of a shallow valley, Figure 5, which sloped downwards in a south-easterly direction. The absolute elevation of the source of gas was about 240 m. Approximately opposite the break, 345 m across the valley (south-westwards) at an elevation of about 232 m was the place, as suggested by Burgess and Zabetakis, at which ignition occurred. The bottom of the valley between these two points was at an absolute elevation of 225 m. The average gradient of the valley bottom ranged from 1.5×10^{-2} to 1.3×10^{-2} m/m. The isolated pockets of anomalous damage referred to above occurred at right-angles to the axis of the valley, but approximately in line with the direction between the source of ignition and the source of propane.

Flixborough presented a different picture. Although the site of the plant itself was practically flat, immediately against the eastern boundary was a gentle ridge which ran north and south for some kilometres. At its steepest, to the north, the ridge rose from 8 m to 61 m at a gradient of 0.19 m/m and thus was a considerable feature of the terrain. From the damage to the surrounding communities it is apparent that explosion pressures must have been considerably polarised towards the north—that is in the general direction of the line of the ridge. However, the line of maximum damage was apparently not parallel to the ridge as communities on top of, and set-back from the top of, the ridge were amongst the most severely affected. It would appear that in the case of Flixborough, terrain had little or no effect on the blast wave. The reason why it was polarised may possibly lie in the process of explosion, already considered.

IX
PREDICTION, PREVENTION AND PROTECTION

Are these words realistic and realisable aims or idle fancies? Prediction–of what, how, and when will it happen? Similar qualifications could be applied to "prevention" and "protection". Plainly, so far as the unconfined vapour cloud is concerned, what it is sought to predict is the occurrence of blast following ignition. Indeed, the ultimate objective would be to predict blast wave pressure levels quantitatively in three dimensions. Different levels of prediction lie below the ultimate objective, however–predictions such as: where will the vapour escape, how much will escape and in what form, what shape will the cloud take prior to ignition and where will it ignite?

Does prevention mean, in this context, preventing the initial escape of vapour, preventing ignition or somehow engineering–given ignition–to prevent blast? "Protection" is defined more straightforwardly–protection is the design and erection of plant and other structures capable of withstanding the predicted blast and capable in turn of protecting personnel within them from serious injury. As matters stand at present, however, it is only the mere act of defining "protection" which is straightforward. It will be clear that proper protection in turn depends upon sound prediction and, as will become clear, upon the form of blast wave which it is *assumed* will arise.

The situation is currently far from satisfactory in any of the three areas considered here and in consequence should be kept constantly under review. The aim of such a review would be to take a realistic attitude to the risks, commensurate with experience, modifying or even changing views which might have been validly held on the basis of previous experience. Naturally, such a flexible approach will only be "right" whilst it is held. What, therefore, is to be done about the majority of operating process plants which, at any time, actually exist, as distinct from those on the drawing board? Plainly a proven approach is correct in those cases; therefore an "organic" programme for dealing with UVCEs must grow on a firm foundation. It is the "firm foundation" which is considered here. I would submit that there are only two bases for such a foundation: either (1) it is the scientific and technical consensus, or (2) it is theoretically demonstrable that the approach is applicable, in all relevant circumstances, with accuracy.

It would, of course, be preferable to have both rolled into one. There are reasons, however, why this is not possible at present. These may already be apparent, but will be set down again shortly. These two statements beg a number of questions, such as: what constitutes a consensus, what are "relevant circumstances", and how accurate? Furthermore, given the indisputable fact that UVCEs can do great damage, would it not be more appropriate that methods of, and for, prediction, prevention and protection–adopted as the foundation of an evolving appraisal–should err conspicuously on the side of safety?

In the following I have endeavoured to set down what appear at present to be the criteria for a realistic appraisal of UVCE behaviour. These are based firstly on the statistics, and secondly, either on conventional TNT equivalents or the blast theory reviewed in the Appendix.

Prediction

The statistics–Table 1 and Section IV–suggest that UVCEs arise more readily from forceful emissions of unsaturated compounds. The combination of these two factors and the occurrence of UVCEs is quantitative in the case of ethylene. Ethylene's higher homologues, however, do not show such quantitative correlation. One reason for this might be their lower laminar burning velocities. Another cogent reason might be their lower process or storage pressures–giving less forceful emissions upon discharge, other things being equal. Both of these factors influence flame speed, which on any view is a fundamental criterion in the development of UVCEs.

Again, in the five cases where it is believed hydrogen or ethylene formed a major portion of the discharging materials, the remainders being intrinsically less aggressive in ordinary combustion terms, blast occurred. The interesting thing here, however, is that hydrogen alone produces relatively weak blast, even under conditions of extreme turbulence (*eg* Case 42). Another material which is reluctant to give rise to UVCE behaviour is methane gase at *normal* temperatures and pressures, although in the form of a vapour at very low temperatures it can cause blast.

But it is *not* the fact that sometimes reactive materials do not cause blast that should occasion surprise; it is rather that often otherwise ordinary flammable materials escaping without great force do cause UVCEs that should be noteworthy. Prediction, naturally, must take account of this factor. Of course, one important aspect has not yet been mentioned, and that is the quantity of flammable material released. This is probably at least as important in determining the *course* of events following ignition as are burning velocity and turbulence. Given the propensity to cause blast (whatever that might be due to in any particular case) then the quantity of flammable involved is of overriding importance so far as the severity of the actual blast is concerned. And yet the "correlation" with quantity of flammable material is obscure–Figures 1 and 2. The best that can be said is that generally the larger escapes are associated with larger explosion yields (in terms of TNT). As this is almost self-evident anyway it is not particularly helpful. Nevertheless, any attempt to identify a trend would, in my view, be wrong because of the significant number of high yields associated with small quantities of flammable. In any event whatever trend (or trends) might be thought to exist must, I suggest, become asymptotic to the 100% yield value as *a priori* there cannot be greater yields than this. Therefore, the problem is much more complex than merely trying to achieve the best straight-line correlation.

To these difficulties must be added the behaviour of hydrogen in Case 42. As discussion of that event (Section V) has revealed—assuming that the quantity of hydrogen involved was only 10 kg, the corollary of which is that the yield is increased to 8%—then this places all the other data into the category of relatively highly inefficient, in blast terms. However, a boundary may be drawn which encloses all the blast data including Case 42 (calculated as 10 kg and 8% yield) if median yields of 25% and 45% respectively are attributed to Cases 49 and 83. This boundary expressed in mathematical terms is approximately:

Yield (%) = 4 x natural logarithm of quantity (kg).

Whatever validity this expression might have, it could not extend to quantities beyond 7.2×10^{10} kg, nor below 1 kg.

The problem with such an expression is that it could be far too onerous to use for design purposes: it vastly overestimates the yield in the majority of cases; it may be based on questionable assessments of TNT yield and no clear indication is given of the meaning of "quantity". In respect of the latter comment, "quantity" should be consistent and, preferably, what was actually consumed during that part of the combustion process which caused blast, *ie*, presumably that fraction of flammable cloud in admixture with air and within the limits of flammability. Further, the expression is purely empirical and has no apparent physical justification; it cannot be related *in any way* to the inherent properties of the flammable material.

Actually this is the problem which besets any correlation based upon TNT equivalents. This is not to say that TNT equivalents are not useful nor that design based upon them will not be adequate. The problems of using TNT equivalents become clear if the leading references are consulted. Brasie and Simpson[107] suggest that TNT yield should be predicted on the basis of a constant fraction of the *total* energy derived from the combustion of the flammable vapour or gas which is produced upon escape. This naturally presupposes proportionately similar explosible fractions of the total escape to exist at the moment of ignition, independent of the size of the escape and for how long it has lasted. This conception is periodically reiterated in articles, *eg* Nelson's[162], which urge UVCE blast protection design.

Brasie and Simpson suggest a TNT yield equivalent of 3% to 4% of the total combustion energy produced by a cloud when it involves hydrocarbons. They suggest a far higher value for hydrogen, 30%, although the precise evidence for this is not clear. The one identifiable incident they do allude to, however, is Jackass Flats (Case 42), where the yield is definitely less than 10%. Nelson suggests 1% to 5%[162], based upon "analysis of actual chemical plant vapour cloud explosions", with the rider that for "establishing separation distances" the value of 2% should be used. Plainly, wide variability is the general rule in this approach.

Let us consider the TNT equivalent approach step by step:

(1) Estimate the quantity of vapour likely to be produced;
(2) Assume that all of this vapour burns explosively (this implies mixture with air in the concentration range between the lower limit and stoichiometric);
(3) Evaluate the combustion energy release of this quantity of vapour;
(4) Divide this energy by the energy released upon the detonation of unit mass of TNT;

(5) Evaluate the actual explosion yield by using the explosion yield factor (3% or 4%–Brasie and Simpson, or 1% to 5%–Nelson).

What is done with the final figure is considered in the sub-section on protection.

An alternative approach is to branch away after step (3) and use the theory set out in the Appendix. Unfortunately, as with TNT equivalents, some assumptions still have to be made, but these can be based upon experience. Flixborough suggests that the value of K in Equation 53 should be 380 $m/s^{4/3}$. Equation 56 then enables the computation of $\dot{p}_c$ (8.8×10^6 Pa s^{-1} for cyclohexane) and use of this value–appropriate to the flammable material involved–in the relevant Equations will generate the p, ϕ and t_o. The complete waveform at all locations may be derived accordingly. K is a parameter of motion. It is not fundamental in any sense and the value suggested above could only be used for hydrocarbons.

Techniques for estimating the rates of escape of materials are straightforward. What is not so clear is how to deal with a developing cloud where its size is strongly time-dependent. Brasie and Simpson suggest that the size of cloud, evaluated for the purpose of TNT equivalence, should be that which will exist 30 seconds after the escape starts. In the case of a dissipating emission (*eg* the sudden discharge of a cryogenic material) the quantity of vapour considered should be the maximum which can occur, irrespective of time–usually within 30 seconds and during, or at the conclusion of, the flash evaporation phase.

All materials capable of forming vapour/air or mist/air mixtures of concentrations above the lower limit of flammability are potentially capable of causing UVCEs. Prediction involves identifying the process or storage risk which presents the maximum hazard and calculating accordingly. The identification of two or more widely separated but significant risks may mean that protection against a UVCE from one will not be sufficient for complete protection. Further, the prevailing wind *must* be taken into account, although it is doubtful whether day-to-day variations in, say, a Disaster Plan could ever accommodate it. So far as the fixed structures are concerned, the blast they are designed to withstand should be the one "focussed" at the point which allows for 30 seconds drift in wind. What can be done in any particular instance will depend upon matters such as annual wind speed and direction data, and the number, type and disposition of significant process risks. Each problem requires separate study.

Prevention

Prevention in this context means prevention of a UVCE, not necessarily prevention of a fire-ball, still less prevention of escape of flammable material. Whilst the latter undoubtedly is the desirable objective, it is largely a matter of assiduous but ordinary operating and engineering competence.

Prevention of ignition may be possible. There are many instances on record[44] where ignition has not occurred. But sources of ignition are legion and those cases of non-ignition of major clouds must be considered to have happened by chance rather than design. Moreover, the emission itself may generate ignition sources, especially if it is forceful. It is right to catalogue these if only to emphasise the point: furnaces; surfaces over 300°C (or even less) including steam pipes; electrostatic discharge; mechanical damage or rupture of electric cables, explosion proof electrical fittings, earth-bonding straps; frictional sparking by flying lagging cladding *etc.* The list is not exhaustive.

Given an escape of vapour, therefore, ignition must be considered likely. What can be done to prevent blast? Will water drencher systems have any effect? Experimental work[138,141] suggests that conventional water sprays at best have no effect; indeed, they would appear to have a promoting effect on combustion. Clearly, however, this is a matter of degree. There must be a point when the combination of spray density and dispersion will extinguish combustion. The criteria for extinction are not as yet precisely quantified, but Vincent[141] alludes to a dispersion pressure of 13.8 MPa (2000 psi), and Eggleston, Henera and Pish[138] suggest a spray density in excess of 1.46×10^{-3} $m^3/s\,m^2$. The power requirement for this level of prevention is prodigious: on an ideal basis it amounts to 20 kW/m^2. Prevention of a UVCE by this method on a site 200 m square for, say, 10 minutes would require both 35 000 m^3 of water and the consumption of at least 1.3×10^5 kW-hr of energy by pumps of at least 800 MW capacity. In Britain, at the ordinary *domestic* price of electricity, running the sprays for just 10 minutes would cost in excess of £3000.

This example is not meant to be rigorous, but it is fairly real. The capital cost would be astronomic. Unfortunately, in this instance, lower investment in a reduced level system might not mean merely "getting what you pay for", as there is the likelihood that a modified (reduced) velocity of sound in the dense dispersion could *enhance* blast effects (see Section VI).

Techniques of prevention by other means may be envisaged. All present problems as difficult as the water spray. It is the case that at this time prevention can be achieved economically and with certainty only by assuring plant integrity.

Protection

UVCEs may be predicted and their effects forecast by two methods. At present they cannot be prevented; given a major escape the probability is that ignition will occur rather more frequently than 14 times in every 15 incidents[44], and that blast will occur on rather more than half of those occasions. Protection is therefore desirable. Unfortunately, like many desirable objectives, it is not easy to achieve. Protection is two-fold: of the process workers and plant and of the surrounding community and its buildings. The point has already been made that most process plants and communities exist now. For historic reasons they have tended to develop side-by-side. Indeed, this has hitherto been encouraged. Only within the last decade has mobility and emancipation of labour enabled employees to reside at greater distances from their places of work. Now other, ecological and environmental, forces are at work and it is likely that cheek-by-jowl developments common in the last 150 years will gradually disappear. Nevertheless, in many respects it is a spontaneous trend which, in decades to come, might reverse, thus re-exposing the community to these risks. In any case it will take many years for a large degree of dispersal to come about.

Because most communities and industries are independent entities—despite their obvious dependence upon each other—their contiguous development can only be prevented by law. This can be the only socially certain method of community protection whatever view is taken of the real risks. Plant and employee protection, however, is largely a commercial exercise. This is certainly true so far as the plant hardware is concerned and not untrue in respect of the employees, although employers in most countries have statutory obligations to fulfil. In Great Britain the Health and Safety at Work *etc* Act of 1974 requires:

> It shall be the duty of every employer to ensure, so far as is reasonably practicable, the health, safety and welfare at work of all his employees

and

> It shall be the duty of every employer to conduct his undertaking in such a way as to ensure, so far as is reasonably practicable, that persons not in his employment who may be affected thereby are not thereby exposed to risks to their health or safety.

It is of interest to note that the onus is on the *employer*. If a community chose to develop, as it can at present, to the very boundary of a plant whose latent hazard–even if of extremely low probability–is assessed as high, it would be the plant Operator who would be responsible for the risk to which the community became exposed–not the community. Whereas this might be considered reasonable by some, it could have the following consequences: the company terminating its plant operations at that location with the possible consequence of additional unemployment in the area; or, and probably more likely, the company adopting the level of protection which "is reasonably practicable" (and which is never totally divorced–even judicially–from commerce) and in reality not achieving much more community protection.

It is the "reasonably practicable" factor which pervades the Health and Safety at Work Act and which determines the level of protection provided by any process Operator for his employees. This is not the place to embark upon the detailed requirements of the Act nor upon reliability data, costs and accident statistics which will form the basis of any assessment of "reasonably practicable". Here the concern is what can be done within a process plant to ameliorate the risks of injury and damage in a UVCE without reference to what is economic or even practical.

Protection for consistency should be expressed in terms compatible with those used in the method adopted for prediction. If TNT equivalents are adopted so explosive, shock-wave, loadings are the appropriate forces for the protected designs to resist. Explosive blast-resistant design is a well developed topic–one volume[163] has a number of papers which deal with it. Another publication[164] contains detailed design data specifically enabling the design of explosion resistant structures. Bradford and Culbertson[17] contains useful references both in the text and in the appended discussion. In this type of design relatively high strength subsists in the inertia of the structure; for ordinary explosive shock loadings the resonance period of most ordinary constructions would vastly exceed the explosive impulse durations. This means that modest additional strength is necessary to accommodate the impulse, but the structure does *not* require to withstand the static load to which the overpressure would be equivalent (simply because the overpressure is not static).

Of course, this approach *assumes* the pressure wave form–*a wave for which there is absolutely no physical evidence whatsoever.* But it is at least a consistent approach. No case is known in which an explosive resistant construction has failed in a UVCE blast. But neither is any case known of explosive resistant construction exposure to UVCE blast.

The conception of UVCE blast expounded in the Appendix implies pressure durations one to two orders of magnitude longer than for conventional explosives releasing similar energies–sufficiently long to be considered essentially statically applied. Whilst this should cause no problems for the civil engineer, all conventional aesthetic considerations would need to be sublimated, as the appearance of the resultant structures would be robust in the extreme. Of course, such structures would also withstand shock waves having the usual explosive profiles. Design based on static loading would err on the side of safety providing blast pressures were correctly forecast.

It is clear that buildings can be built to withstand blast. It is desirable, if these buildings are built and are designed to accommodate personnel, that they also protect persons within them. This requires keen attention to detail, in particular to windows, doorways, ventilation openings, service openings *etc.* This is, of course, true whatever view is taken of the likely blast wave pressure profile and duration. All openings and glazed areas should seal during a UVCE and should be of no less strength to pressure than the structure itself. They should be capable of moving with the structure and accommodating the resulting strain without over-stressing. Quite apart from the hazard of flying fragments of glass *etc*, the tornado-force winds[18,165], which would enter through any breach in the walls, could be gravely dangerous. Finally, self-contained atmospheres within buildings occupied by personnel would be essential to eliminate the possibility of invasion of these atmospheres by hot and probably toxic combustion products.

There are no new concepts involved in protection within process plant. As this is not a text on civil engineering, details of the techniques involved are not appropriate here. However, the literature[17,164] provides a starting point for explosive-resistant construction. If the view is taken that the TNT equivalent assessment is uncertain, does not go far enough or indeed is wrong, then quite straightforward civil engineering will enable the design of structures of any reasonable process size to withstand statically applied pressures. The pressures involved will be inversely related to the distances of the structures from the explosion centres. A degree of compromise exists, therefore, in most practical situations.

X
SUMMARY AND CONCLUSIONS

1. In the context of this report an Unconfined Vapour Cloud Explosion is the result of the ignition and subsequent burning of an accidental release into the ambient atmosphere of gas, vapour, mist and/or dust. The principal characteristic of the event is the creation of damaging pressure forces within and beyond the boundary of the flammable cloud. The meaning of the term "unconfined" does not exclude proximity to or envelopment of natural or man-made structures, specifically including process plant, nor proximity to the ground.

2. The experience of UVCEs is limited, although they are now apparently happening at a rate of two to eight *per annum* (Table 1) throughout the world. The recent experience of all vapour cloud ignitions capable, in principle, of causing blast is that these arise at a rate of three to eleven *per annum.* The statistics reveal that few major chemical, petrochemical, process, or petroleum plants have escaped the possibility or the actuality of an UVCE. The various possible means of transportation also give rise to the risk.

3. Since 1921 and to February 1977, there have been at least 100 incidents potentially capable of involving quantities of flammable material of approximately road tanker proportion or greater. Many of the early incidents involved liquefied petroleum gases—propane and butane—in transit. The *frequency* of these has lessened, but only slightly. Added to these numbers now are many more static process plant failures, the reasons for which are less overtly the error of the "man in charge" than seems to be the case for LPG. Of the 100 incidents, 56 gave rise to blast, and of a total of at least 725 fatalities, 385 (53%) died in events involving blast. More than half of all the incidents happened in the last 12 years, and 60% of these were associated with blast. 58% of all who died in the incidents considered in Table 1 died in the UVCEs which have occurred in the last 12 years.

The trends in frequency, proportion of incidents producing blast, and fatalities in UVCEs, based on these figures, are all upwards.

4. The statistics of the flammable materials involved show that they have a very wide range of physical, chemical and combustion properties. The only common factors would appear to be that the materials must be flammable and must be readily able to form mixtures with air within the appropriate flammable ranges. There would

also appear to be a distinct correlation between the occurrence of UVCEs and the existence of molecular strain within the molecule of the flammable material. The involvement of hydrogen, as a component of a flammable mixture, seems to be associated quantitatively with blast, although pure hydrogen does not cause blasts of expected magnitudes. On the other hand, methane, when cold, can cause blast, although this apparently does not happen with the gas at ordinary temperatures.

No flammable–or even combustible–material can be excluded as a potential "fuel" for a UVCE if the conditions of storage or processing are such that a large escape could mix rapidly with air.

There would appear to be a clear statistical correlation between the involvement of materials whose molecular structures are "unsaturated" or cyclic, and the occurrence of blast.

5. Hitherto, UVCE damage correlations have been based upon their equivalence to a conventional high explosive–TNT. The yield of the explosion in terms of the weight of TNT which would cause *similar* damage is evaluated. The weight of flammable material involved is also evaluated–either as a total weight or in admixture with air and within the flammability limits. The energy released on detonation of the equivalent quantity of TNT is then set in ratio to the energy released by combustion of the flammable material and the result expressed as a percentage. This percentage is referred to as the "efficiency" of the blast.

In only 22 of the 56 UVCEs has sufficient information been available to enable computation of blast efficiency. Plotting the blast efficiency against quantity of flammable involved and against a composite (but arbitrarily constructed) combustion parameter, Figures 1 and 2, has failed to disclose any realistic correlation.

In no case, based upon the current data, can a blast efficiency (related to TNT) be predicted with any confidence whatsoever from a knowledge of the flammable material and the likely quantity escaping into the open air.

6. With the exception of hydrogen, it would appear to be necessary that most flammable gases or vapours capable of causing UVCEs, at the point of escape as well as in explosible mixtures with air, display a density greater than that of the ambient atmosphere. Such mixtures naturally tend to form low lying, two-dimensional clouds. No specific correlation seems to exist between the TNT equivalent "efficiency" of a UVCE or indeed even the occurrence of a UVCE and the physical state, disposition, orientation or force of emission of the flammable material, although UVCEs are associated more frequently with flashing "superheated" liquids rather than with gaseous or flashing cryogenic escapes.

The delay prior to ignition, from the point in time when emission commenced, is also extremely variable; smaller emissions are generally associated with shorter delays, although large escapes show sometimes short, sometimes extended, ignition delays. Clearly, here the rate of dissipation is an important but, on present information, unquantifiable factor in interpreting the data.

Whilst it is possible in principle to envisage a vapour cloud developing in the presence of a source of ignition such as a flame, the whole body of evidence suggests that it is unlikely. The more usual occurrence is the fire-ball or boiling-liquid expanding-vapour explosion (BLEVE), which is both spectacular and damaging, but generally not associated with more than very mild and local pressure effects.

7. All theoretical assessments of UVCEs require that the explosion flame, as it propagates through the flammable cloud, should accelerate to speeds of the order of that of sound in the unburnt mixture. Such speeds are two to three orders of magnitude higher than the fundamental laminar burning velocities of, for example, hydrocarbons. Such enhancements of flame speed cannot be explained solely in terms of turbulence, although turbulence undoubtedly exists and does assist flame acceleration. These speeds do not imply detonation although detonation is considered to be a possibility in UVCEs. However, substantial further acceleration would be necessary to achieve fully developed detonation.

It is considered significant that hydrogen with an inherently high burning velocity should, seemingly alone amongst those gases or vapours less dense than air, cause blast, and yet nevertheless apparently not achieve high blast efficiencies (on the basis of TNT equivalents). Indeed, in the only well documented incident—in a technical sense—involving pure hydrogen, most of the flammable volume formed did not burn. An explanation based upon "aerodynamic" influences can explain why hydrogen does not burn aggressively, why warm methane gas does not give rise to UVCEs and why materials of higher molecular weight, or higher density or both, than air should exhibit accelerating flames.

8. An elementary theory of UVCEs has been set out in the Appendix to this report. This seeks to quantify pressure effects at *all* locations whilst at the same time remaining compatible with observed behaviour. The theory considers UVCEs as overlapping two-stage events: energy release followed by energy transport. No assumptions have been made about the mode(s) of energy release, although the model assumes hemispherical symmetry. Energy conservation is assumed in the energy transport phase of UVCEs.

The theory is not fundamental—in the sense that it is not based upon gas-dynamic first principles; it seeks merely to explain observations and effects and be self-consistent. It explains factors such as: the failure of those close to UVCEs to hear the explosion; the apparently uniform pressurisation suffered by the boundaries of enclosed spaces independent of their position relative to the "centre" of blast; the widely varying TNT equivalent "efficiencies" found for different UVCEs—efficiency being, on the basis of this theory, a function of many independent parameters. The theory predicts long duration pressure waves which would considerably exceed the vibrational period of most structures and thus manifest as static loadings.

Whilst it is based upon hemispherical symmetry, it makes no assumptions about the shape, form or direction of the flame. Naturally, this in reality would be expected to generate its own local pressure effects. Nevertheless, the theoretical pressure profile adopted resembles fairly closely profiles anticipated by those who have considered symmetrical flame movement, in a hemispheric system, from a central point of ignition. Whereas hemispherical symmetry has been adopted as a convenience, other symmetries or indeed asymmetric vapour clouds could undoubtedly be appraised by computational techniques. Notwithstanding the symmetry used in deriving the theory, its most informative application is to unsymmetrical, or uniaxial, or two-dimensional clouds. Specifically, the theory predicts "aerodynamic" flame accelerations for flammable concentrations in air displaying a greater density than the ambient atmosphere. It predicts the converse, flame deceleration, for those mixtures manifesting lower densities than the surrounding air. With hydrogen the effect could lead to isolation of the flame from the main body of the flammable cloud–the "pinch effect".

9. TNT equivalence has led to the idea that UVCEs produce blast waves having essentially the same form and proportions as explosive shock waves. There are absolutely no grounds whatsoever for such similarity, either in conception or in reality. Theory predicts that a large pressure wave will tend to steepen on the leading edge and become more shallow on the trailing edge. A uniform sinusoidal profile, for example, will distort progressively towards a saw-tooth shape, but it is most unlikely that it will ever transform totally, within the range in which it can do damage. The difficulty this gives rise to is that, whereas the pressure levels from a UVCE and its "equivalent" TNT counterpart may be similar, the energy content of the UVCE pressure wave may be many times greater (reflecting the intrinsic energy differences between TNT detonation and vapour combustion). The only way this additional energy can appear, unless it disappears or somehow is not released, is in extended pressure waves–necessarily of long duration.

TNT equivalents are recognised as inadequate close to the focus of blast, either diagnostically or predictively. They are likely also to remain inadequate remote from the blast centre. Whereas this is of little absolute importance, although unfortunate, in diagnostics, it is of profound importance and of great concern in prediction, especially in the planning and design of structures. The problem is that structures designed for a given explosive shock wave overpressure and duration would probably collapse if the duration of overpressure were extended into the static loading régime as, it appears, might be the case.

Reappraisal of the possible durations of pressure loading in UVCEs is essential now in view of the current design studies being undertaken by architects.

10. Investigation has been adequate following most UVCEs. Only latterly has it become the rule to collect comprehensive data and publish it in accessible journals. Comprehensive data is required before the right diagnosis can be reached and the right diagnosis is essential before any reliable future predictions can be made. Only reliable predictions enable adequate protection to be designed.

As yet there has been no incident about which certainty exists regarding all the following factors. It is the strong recommendation of this report that in all future incidents these aspects be as comprehensively recorded as may be possible:

(a) Flammable material which escaped;
(b) The exact point from which the material issued to atmosphere;
(c) The orientation of the escape, its elevation and form;
(d) The volume and mass rate of escape of the material and its state–both as functions of time;
(e) The weather conditions at that time;

(f) The delay prior to ignition of the "vapour" cloud and its form at the moment of ignition;
(g) An accurate description of the mode of flame propagation and all perturbations and anomalies;
(h) Other factors relating to cause, *ie*, fire observed *before* the vapour cloud developed or exploded;
(i) A comprehensive pattern of damage, or non-damage, correlated with distance and catalogued according to either the minimum pressure or impulse necessary to do the damage or the respective maxima which would just not cause damage; particularly in relation to objects deformed by drag forces.

11. *It is recommended that diagnostic techniques be extended using imagination and a strong analytical but multidisciplinary base, with the object of extracting as much latent information as possible from both adequately recorded past UVCEs and those to come.*

12. If the "pinch effect", referred to above, is responsible in part for flame acceleration, theory suggests that it cannot arise unless: the mixture which burns is more dense than the surrounding atmosphere—as already mentioned; the mixture is burning from a point of ignition at an extremity of the cloud; and the shape of the cloud into which the flame is propagating is conical or wedge-like (with the cone or wedge semi-angles less than 90°) and getting thicker in the direction of propagation.

Specifically, this means that experiments with symmetrical clouds of hemispherical shape, whether ignited at the centre, at the outer boundary or anywhere between, will be unlikely to display the flame accelerations necessary to model UVCE behaviour. This is confirmed by all the "large-scale" well documented experimental work to date.

It is suggested that edge ignition of dense unidimensional or two-dimensional explosible clouds be studied in the future. Also a study of the effects of non-homogeneities might be profitable—in particular the influence of the presence of high concentrations of droplets in the cloud and the interactions of burning vapour clouds with dense process plant, buildings or vegetation.

Theoretically, the influences of high concentrations of non-compressible objects in dispersion in a burning cloud could be significant in the development of high pressures—this theory in its predictions does not differentiate between solids and liquids in gas, nor does it matter, apparently, how large the non-compressible items are—so long as they form a significant volume fraction.

13. Mortality associated with UVCEs arises through blast and fire. The lethal element of blast is associated with the collapse of buildings and projection of fragments, *not* with direct blast effects. Current estimates suggest that maximum rates of pressure rise even at the centres of UVCEs are less than will cause critical injury. Within the burning cloud, but away from buildings, deaths from burn injuries are usual.

The mortality data discussed in this report does not differentiate between blast and burn fatalities. The data is, of course, based upon past incidents in unprotected plant and is correlated with accidents involving conventional explosives. The correlations are compelling and indicate, in the absence of protection, the existence of two possible extremes of mortality (low and high), depending upon whether plant is in rural or urban surroundings.

The adoption of appropriate protection measures would materially affect the levels of mortality and are strongly advocated.

14. What constitutes appropriate protection depends upon the view taken concerning the basic mechanisms involved. One consistent approach, which assumes the blast wave-form, is the TNT equivalent method. An alternative is the energy conservation method developed in this report. The two are not compatible and buildings designed according to the former may be seriously damaged if the blast actually corresponds more with the latter. Both approaches, however, present few problems, as techniques of building construction can deal with either form of blast wave.

It is important, however, to reflect that "protection" is not worth anything unless it protects; that is, unless lives are saved, injuries made fewer and less serious, and unless plant losses, both material and consequential, are reduced. *It is recommended that a programme of study be instituted to consider the implications of blast duration on the design and fabrication of protected structures and to evaluate from first principles, from well documented UVCEs, and from experiment the likely duration of UVCE blast pressures.*

In future some protected structures will be used by personnel—they may even come to be considered as refuges in disaster situations. If they have been designed and constructed properly, they will withstand the UVCE which has been forecast. This alone, however, may not be sufficient to prevent injury to persons within: not only has the structure to survive—so have all its components, in particular windows and doors, which must neither break nor open during blast; further, the means of ventilation or air conditioning adopted in the building has to prevent ingress at high velocity of the hot and toxic external atmosphere. Ingenuity might in due course enable secure refuges to be provided for personnel whose jobs require them to go into process plant.

15. Protection, to be reliable, requires that the UVCE, from the effects of which protection is required, be realistically predicted. It is clear that the maximum credible event in any chemical plant can be predicted. It is plain that a building between two or more areas of risk has to be assessed with regard to them all.

What is more difficult is allowing for the effects of wind. Whilst allowance can be made in statistical terms, the chances of the statistically likely wind at the time of an actual UVCE may be low or even negligible, depending upon the climatic character of the particular area.

Another aspect is the polarisation of blast. Why or how this should happen is a puzzle. It can be explained in terms of the theory developed in the Appendix to this report; as being due to asymmetry and flame acceleration influences—the "pinch effect" already referred to could act not only as a flame accelerator but also as a flame guide (or "gun"). If so, the position of the point of ignition in relation to the point of escape (coupled with the wind velocity) is relevant. This polarisation, which may well be intrinsic to the UVCE, is to be distinguished from polarisation caused by meteorological conditions, which can lead to isolated pockets of locally anomalous damage in areas and at distances where such effects should not otherwise occur. Again, protection against such damage may not be possible, for whilst it seems that the effects can be adequately explained, even forecast, they result from atmospheric behaviour as yet not under our control.

16. The only certain method of protection is to prevent UVCEs happening, by which I mean preventing vapour escapes. Unfortunately, whilst materials and personnel are generally much better than average in chemical plant, neither is perfect all the time. Materials are only as good as the men that specify them, that develop them, that manufacture them, that test them. Personnel are as good as their self-discipline, their knowledge, their training and their intelligence. Whilst the world depends on men, mistakes will occur, most of which will be trivial in all respects. A few, however, will be serious and amongst these will be UVCEs. Published statistics reveal their frequency and the corresponding financial losses. Prevention, within the present framework of a large-scale, low-profit margin philosophy, can be no better than in the past; indeed, the trend is to worse and more frequent incidents.

Despite this somewhat pessimistic view, techniques are being developed which by their very nature will be slow to show their effects, largely because if they work incidents will not happen. These techniques should be well known by now and are discussed at length and in detail elsewhere. They form no part of this study, but it is right that I should at least name them. They form the core of the current insurance *maximum credible loss* philosophy. A similar exercise, although with a more stringent and technical basis, is soon to be required in industry in Britain, where plant will be judged in major chemical hazard terms, by the inventories of hazardous chemicals which will be required to be notified to the Health and Safety Executive. Warning of the intention of the Health and Safety Commission to require notification of inventories of hazardous chemicals beyond certain minimum levels was given in the First Report of the Advisory Committee on Major Hazards, published by HM Stationery Office in 1976.

Loss prevention techniques require the unusual combination in individuals or groups of individuals of common sense and imagination. Certain aspects of the problem are easy—identifying plant and processes which represent the major risks, for example. What is not so easy is identifying points of hazard, prospective areas of failure and their consequences.

Having identified the hazards and the likely consequences, it is necessary then to institute a system of rigorous controls over operations, inventories, maintenance, construction, supervision and inspection. This is what is done routinely now by many major companies, which use sophisticated procedures such as fault-tree analysis, minimal inventory processes, reliability studies, planned maintenance *etc.*

An alternative to preventing emissions of vapour is to attempt to ensure that such emissions do not give rise to blast. Whilst the vapour cloud would probably ignite, damage could, in principle, be restricted to the plant itself and not the surrounding community. Indeed, it is likely that damage within the plant would be much less severe and much less extensive. The statistics suggest that the chances of survival of personnel on the sites would also be higher. Unfortunately, one technique of blast prevention (rather than escape or ignition prevention), whilst possible, is, it would appear, scarcely practicable—involving huge capital costs and massive running costs. Careful consideration suggests that less than wholehearted adoption of this form of prevention (water spray) might lead at best to no effect and at worst to an enhancement of blast.

APPENDIX

Vapour Cloud Formation

This text is not directly concerned with the details of the development of the flammable cloud. Nevertheless, it is a necessary concomitant and predecessor of the UVCE and to ignore the broader aspects of cloud development would be wrong. Moreover, one or more of the various ways in which flammable clouds can form *may* predispose them to produce significant blast effects. The conditional tense is emphasised because what appears later strongly indicates, I suggest, that other factors are more important.

For a vapour cloud of "significant" size to form requires that the forces leading to its creation override dissipative agencies. By far the most important dissipation mechanism is atmospheric turbulence which is related to wind speed and other factors. Broadly, the greater the degree of "override" the larger the resulting flammable cloud will be (given sufficient flammable material to start with). The forces leading to the creation of a vapour cloud are derived either from energy contained within plant or energy contained within the flammable material (its internal energy) or both, or from the relative energy excess to be found in the surroundings. The principal forms of energy in this context are pressure and chemically or thermally derived energy. Generally, only the former and the latter can operate independently.

Restricting our attention to the flammables in the gaseous or liquid states–rapid combustion of these occurs only in the gas phase; therefore, for coherent liquids the vapour pressure has profound significance. Vapour pressure has less relevance, however, when considering the behaviour of liquid in a finely dispersed state. Recognition of these factors enables subdivision of emission into the following classes:

1. Gas
2. Liquid above its atmospheric boiling point
3. Liquid below its atmospheric boiling point and
4. Liquid boiling below atmospheric temperature.

The force of emission in each of these classes can be increased by externally imposed pressure, but only those materials falling into class 2 inherently exhibit pressure. Of course, most gases in store (as gases) or in process are under pressure. Categories 3 and 4, however, in bulk (storage) do not require to be pressurised–although in process they may be. Nevertheless, often they are under slight pressure of purging or inerting gases or, in the case of cryogenic liquids (class 4), under vapour pressure slightly in excess of one atmosphere.

The least hazardous of these in relative terms is class 3. It is also the one most affected by atmospheric conditions–largely because, in general, rates of dissipation of flammable (explosible) mixtures compete effectively with rates of generation. Pasquill[166] studied and quantified the problem of evaporation of liquids exhibiting free surfaces and later presented the subsequent behaviour of their vapours as a function of atmospheric turbulence[167]. This he did by adapting Sutton's classical work[168] on free atmospheric turbulence–itself refined over the years by Sutton[169]. One of the deficiencies of this treatment has been that its predictions were valid (with accuracy) only for emissions of zero buoyancy relative to the atmosphere. Recent large-scale testing work[130] has enabled a better refinement of these predictions by more closely circumscribing the values of the appropriate "diffusion" coefficients. Recent papers[170] have highlighted the changes which have resulted.

Whilst the above comments have been related directly to emission class 3, they are also appropriate to the behaviour of *parts* of the emission classes 2 and 4; in particular, those parts which remain in a coherent liquid form after initial flash evaporation. In case 2 the residual coherent liquid may exhibit a boiling point above, or below, the temperature of the surroundings. If the former, then its rate of evaporation will be comparable to that of a cryogenic liquid (4); if the latter, then behaviour will be as for class 3. In classes 2 and 4 two distinct phases of "vapour" formation exist. The first involves rapid, almost explosive (in the physical sense), volatilisation, due to flash evaporation (class 2), or violent boiling on contact of the liquid with its relatively hot surroundings (class 4), followed in both cases by pool-type evaporation which proceeds at much lower rates. The relative hazards associated with each phase are quite different and are easily distinguishable both in conception and in experiment[136]. The rapid evaporation periods of behaviour of classes 2 and 4 will be considered below.

Pure gas emission probably–again in relative terms–represents the next higher hazard class to class 3 emissions. Because of the relatively low density of even highly compressed gases, even the highest capacity high-pressure plant contains only small quantities of gas. At the other extreme large-capacity, low-pressure vessels exist, but rarely fail catastrophically (the dirigible ZR-2, case 1, being the largest recorded case in this class). Failures of a non-catastrophic nature occur more frequently, but, because of the low pressure, do not generate large volumes of explosible mixtures.

Marshall[100] has considered emissions of gas having appreciable buoyancy or momentum and extends the work of Long[171] by generating equations which enable the prediction of hazard distances, flammable volumes of plumes or jets and the masses of flammable material involved He greatly refines and extends methods used previously[21]. This work is considered sufficiently important to quote his final equations. These represent the maximum hazard situation–as no dissipation by atmospheric turbulence is considered.

The equations for momentum jets are:

(1) axial length to the lower limit concentration c_l

$$z_l = \frac{K_1 d_o}{c_l}\left(\frac{M_a T_a}{M_o T_o}\right)^{1/2} \qquad \ldots\ldots\ldots\ldots\ldots\ldots \qquad (1)$$

(2) volume of flammable (explosible) mixture within the flammability limits

$$V_{Fl} = \frac{\pi K_1^3}{9K_2^2} \left(\frac{M_a T_a}{M_o T_o}\right)^{3/2} d_o^3 \left(\frac{1}{c_l^3} - \frac{1}{c_u^3}\right) \quad \ldots \quad (2)$$

(3) weight of flammable material in V_{Fl}

$$Q_{Fl} = \frac{\pi K_1^3 273}{6K_2^2 22.4} \left(\frac{M_a}{T_o}\right)^{3/2} \left(\frac{T_a}{M_o}\right)^{1/2} d_o^3 \left(\frac{1}{c_l^2} - \frac{1}{c_u^2}\right) \quad \ldots \quad (3)$$

where the recommended values of K_1 and K_2 are respectively 9 and 12.7 and the units are SI. Accuracy of predictions using these equations is probably better than 20% and predictions are only this inaccurate for gases of extreme molecular weight relative to air.

The corresponding buoyant phase equations are:

(1) $$z_l = \frac{K_3^{3/5} \dot{V}_o^{2/5}}{g^{1/5} c_l^{3/5}} \left(\frac{M_a T_o}{M_a T_o - M_o T_a}\right)^{1/5} \quad \ldots \quad (4)$$

(2) $$V_{Fl} = \frac{5\pi K_3^{9/5} \dot{V}_o^{6/5}}{27 K_4^2 g^{3/5}} \left(\frac{M_a T_o}{M_a T_o - M_o T_a}\right)^{3/5} \left(\frac{1}{c_l^{9/5}} - \frac{1}{c_u^{9/5}}\right) \quad \ldots \quad (5)$$

(3) $$Q_{Fl} = \frac{5\pi 273 M_o K_3^{9/5} \dot{V}_o^{6/5}}{12\ 22\ 4 T_a K_4^2 g^{3/5}} \left(\frac{M_a T_o}{M_a T_o - M_o T_a}\right)^{3/5} \left(\frac{1}{c_l^{4/5}} - \frac{1}{c_u^{4/5}}\right) \quad \ldots \quad (6)$$

The accuracy of these equations depends upon to what extent buoyancy is intrinsic or extrinsic, *ie* whether it is due to molecular weight or temperature. The suggested preferred values of the constants K_3 and K_4 are respectively 17 and 13.

In general, momentum jets transform into plumes at some distance from their point of origin, depending upon the relative buoyancy of the original gas. This transformation does not occur precisely at one plane but over a somewhat indeterminate space located approximately at a distance from the origin given by:

$$z_{tr} = K_5 w_o \left(\frac{d_o}{g}\right)^{1/2} \left(\frac{M_a T_o}{M_a T_o - M_o T_a}\right)^{1/2} \quad \ldots \quad (7)$$

K_5 is variable for the reason given above and depends strongly upon whether the gas forming the jet is positively or negatively buoyant. Marshall suggests values and himself uses $K_5 = 1.55$.

Marshall accepts that this theory applies to jets issuing at up to sonic velocity. He does not state that this approach is wholly inappropriate for supersonic discharge. Indeed, one reason why he does not may be that in plant situations the accidental creation of an orifice is most unlikely to produce a converging/diverging nozzle of the type necessary, in turn, to produce a highly supersonic flow stream; he makes no reference to it because it is not a real possibility.

The momentum jet expressions given above may be extended to two-phase emissions if any appropriate molecular weight can be assigned to the discharging fluid. This is illustrated below.

Earlier it was stated that in emission classes 2 and 4 two distinct discharge régimes existed: a flash evaporation régime and a pool-type evaporation régime. Whilst it is clear that the respective proportions of the emission which enter the vapour phase in either régime are largely determined by the inherent properties of and imposed conditions on the materials involved and also that pool evaporation and flash evaporation are, or may well be, simultaneous in class 2, but must be consecutive in class 4, it does appear that there is no recorded incident of a UVCE which has resulted *solely* from vapour accumulation from an evaporating pool of liquid. It has always been the vapour derived from the initial flash evaporation which has been the problem.

The evaporation of cryogenic liquids has been studied theoretically and practically[136, 172–182]. It has long been recognised that, when a spill occurs, initially there are extremely violent effects due to the relatively large temperature difference between the liquid and the ground. Not only does vapour form rapidly, but it carries with it a large proportion of liquid. As the ground chills, however, boiling becomes less vigorous until, finally, the liquid surface, although possibly still releasing vapour in quantity, is quiescent.

Clancey[183] has produced equations based on normal heat-transfer criteria for this difficult rapidly-boiling phase. He gives a theoretically derived expression for the weight of liquid evaporated per unit area in one minute.

$$Q_1 = 1.42 \times 10^{-4} c_s \rho_s (T_s - T_b)^2 / \lambda_B \text{ kg m}^{-2} \quad (8)$$

The point is made that the numerical coefficient is greatly dependent on the nature of the substrate.

Japanese experimental work[136] on ethylene gave the following formula for the evaporation rate:

$$Q = 1.766 \times 10^1 \exp[-0.0507t] \text{ kg s}^{-1} \quad (9)$$

or, in one minute, per square meter superficial area:

$$Q_1 = 67.5 \text{ kg}$$

$$\text{or } Q_1 = 0.5035 (T_s - T_b)^2 / \lambda_B \text{ kg m}^{-2}$$

In the Japanese tests the bottom of the basin was formed by pebbles having a specific heat and density respectively of 0.25 kcal/kg°C and 2500 kg/m^{-3}, or, in the same terms as equation (8):

$$Q_1 = 8.056 \times 10^{-4} c_s \rho_s (T_s - T_b)^2 / \lambda_b \text{ kg m}^{-2}$$

As the pebbles in the tests were approximately 20 mm in diameter and formed a bed 0.1 m deep having an estimated void fraction of 50%, the actual surface area per unit superficial area would have been 60 m^2/m^2. Q_1 on an *actual* area basis, therefore, would have been

$$1.343 \times 10^{-5} c_s \rho_s (T_s - T_b)^2 / \lambda_b \text{ kg m}^{-2} \quad (10)$$

Although the situation envisaged by Clancey is different from the test conditions, it is clear that the theoretically predicted evaporation rate is enormously greater than the rate found in practice. Examination of the test work reveals that the rapid boiling phase occupied about 10 seconds only and the liquid lost from the bunded enclosure in that time had not been included in equation (9); in that 10 seconds about 293 kg escaped from the bund. Clancey predicts—equation (8)—that over 1/6 of one minute between 9.7 kg (superficial area) and 584 kg (actual area) would be lost by violent boiling, which is, in the circumstances, reasonably in agreement with experiment.

Class 2 emissions create vapour by yielding their sensible heat excess, over and above the atmospheric saturation condition, as latent heat of evaporation. There is reason to think that, as a class 2 emission moves into regions of lower pressure, it is in near instantaneous equilibrium at all points of flow. As such an emission flows, clearly it becomes compressible at some stage (although never wholly gaseous).

The vapour cloud generated by such emissions takes a form appropriate to the mode of release. Catastrophic and total failure of a full tank produces a different situation from orifice-type discharge, such as occurred at Flixborough and Beek. In the former case discharge is three dimensional with concomitant difficulty in effectively mixing with air, whereas in the latter case mixing is actively promoted and is always efficiently accomplished.

A liquid in process or storage which is supersaturated with respect to the external atmosphere will flash as it becomes decompressed. As it flashes its temperature falls and so does its saturation pressure. Consequently, the driving force behind the emission falls. The rate of flow of a compressible medium through an orifice is dependent upon the velocity of propagation of a pressure wave through it. An orifice flow velocity equal to the pressure wave velocity provides the limiting discharge-rate criterion. Critical two-phase flow is well understood theoretically and the theory is experimentally attested[146, 147, 184–188].

The basic equation for the velocity of propagation of an isothermal pressure wave in a two-phase system (one of which is compressible) is:

$$c = \left[P/\rho \left\{ 1 - \frac{\rho}{(1+\mu)\rho_L} \right\} \right]^{1/2} \quad \ldots\ldots \quad (11)$$

where μ is the mass ratio gas/liquid and ρ is defined (in this context) by

$$\frac{1+\mu}{\rho} = \frac{\mu}{\rho_V} + \frac{1}{\rho_L} \quad \ldots\ldots \quad (12)$$

μ, being a ratio, extends from 0 (liquid only) to infinity (gas only). Minima or maxima can be determined by differentiating equation (11).

$$\frac{dc}{d\mu} = \left[P/4\rho_V \rho_L \mu^{3/2} (1+\mu)^{3/2} \right]^{1/2} \left[\mu(\rho_L - 2\rho_V) - \rho_V \right] \quad \ldots\ldots \quad (13)$$

and equating to zero. Solutions exist when

$$\mu = \frac{\rho_V}{\rho_L - 2\rho_V} \quad \text{and when } \mu = \infty$$

Now $\dfrac{dc}{d\mu} = \infty$ when $\mu = 0$

and when $\mu = \infty \quad c = [P/\rho_V]^{1/2}$

which is the isothermal velocity of sound in a gas.
When

$$\mu_{min} = \frac{\rho_V}{\rho_L - 2\rho_V}, \quad c_{min} = \left[\frac{4P(\rho_L - \rho_V)}{\rho_L^2}\right]^{1/2} \quad \dots\dots\dots\dots\dots\dots \quad (14)$$

For air and water at NTP the minimum occurs when $\mu \cong 1.3 \times 10^{-3}$ —mass ratio (or unity volume ratio) and the corresponding propagation velocity is 20 m s^{-1}.

More useful in this context is the mass rate of discharge of such a compressible two phase fluid $c\rho$.

$$c\rho = \left[\frac{P\rho_V(1 + \mu)}{\mu}\right]^{1/2} \quad \dots\dots\dots\dots\dots\dots\dots\dots\dots\dots \quad (15)$$

whence

$$\frac{d(c\rho)}{d\mu} = \frac{1}{\mu}\left[P\rho_V/4\mu(1 + \mu)\right]^{1/2} \quad \dots\dots\dots\dots\dots\dots\dots\dots \quad (16)$$

which is a minimum when $\mu = \infty$. In other words, the minimum mass rate of discharge of any orifice is only achieved when the fluid flowing is a pure gas or vapour. The mass rate maxima increase monotonically as the liquid burden in the vapour rises.

If it can be assumed that at the orifice the plant fluid is decompressed to one atmosphere and if the initial conditions are known, μ can be evaluated from

$$\mu = \left\{1 - \exp\left[-\frac{c_L}{\lambda}\left(T_L - T_o\right)\right]\right\} \quad \dots\dots\dots\dots\dots\dots\dots\dots \quad (17)$$

μ is likely to be invarient during the forceful part of the discharge, but P does change. P is, however, a function of $c\rho$ only and may be found analytically or by trial and error methods.

The effect created by such a discharge is to powerfully induce air into admixture. in accordance with momentum jet behaviour set out above. M_o, the molecular weight of the fluid at the orifice, however, has to be estimated. M_o is an effective molecular weight and may be determined from:

$$M_o = \frac{M_v\,\rho_V(1 + \mu)}{(\mu\rho_L + \rho_V)} \quad \dots\dots\dots\dots\dots\dots\dots\dots\dots\dots \quad (18)$$

The use of M_o, derived from this relationship, in Marshall's equations[100] would be wrong except within a few diameters of the point of discharge. Most liquids when discharged in this way totally evaporate very quickly and the behaviour of the jet rapidly becomes normal. Probably the error introduced by using the jet equations cited earlier together with $M_o = M_v$ would not be too great.

Blast Waves

Pressure waves capable of causing damage and yet not exhibiting a discontinuity are herein termed "blast waves" as a means of differentiation. Generally, however, no such distinction is made in common parlance. Blast waves are fundamentally unstable in that they quickly degenerate into shock waves. Bradley[40] has given an expression for the time taken for a shock wave to appear:

$$t_s = -\frac{2}{(\mathrm{d}p/\mathrm{d}r)_{max}}\left[\frac{(\gamma - 1)}{2} + 1\right]^{-1} \quad \text{.......} \quad (19)$$

As this is not dimensionally consistent the multiplier $a\rho_o$ is introduced. Further, as the physical properties with which we are concerned are those of the atmosphere, this expression can be simplified to, in consistent SI units:

$$t_s = -\frac{710}{(\mathrm{d}p/\mathrm{d}r)_{max}} \text{ seconds} \quad \text{.......} \quad (20)$$

The use of this expression is one way of determining the pressure gradient near the centre of the explosion, if sufficient witnesses are available to discriminate with accuracy at what distance the shock wave starts to appear.

The pressure gradient can be obtained another way, however, by energy considerations and making assumptions regarding the nature of the atmospheric pressure wave-form. The assumed wave-form *may* have no strict validity and its only justification is that it might be generally useful in predicting blast effects in the near and far fields. It is considered that certain criteria are essential in representing the actual profile of the *initial* pressure wave:

(1) The pressure gradient is uniform throughout and shows no discontinuities;
(2) The pressure gradient at the boundary of the pressure wave is zero and
(3) The pressure gradient at the centre of the explosion is zero.

One of the simplest forms of pressure wave which satisfies these criteria is:

$$p = \frac{p_c t}{2}\left[1 + \cos\left(\frac{\pi r}{at}\right)\right] \quad \text{.......} \quad (21)$$

This describes the pressure at all points within an envelope bounded by the radius to which a pressure wave would have propagated in time t. The expression has no meaning when $r < o$ or $r > at$. The form of curve defined by this expression is the "companion to the cycloid". The expression is appropriate only to the energy liberating phase of the phenomenon. Figure 12 illustrates its form.

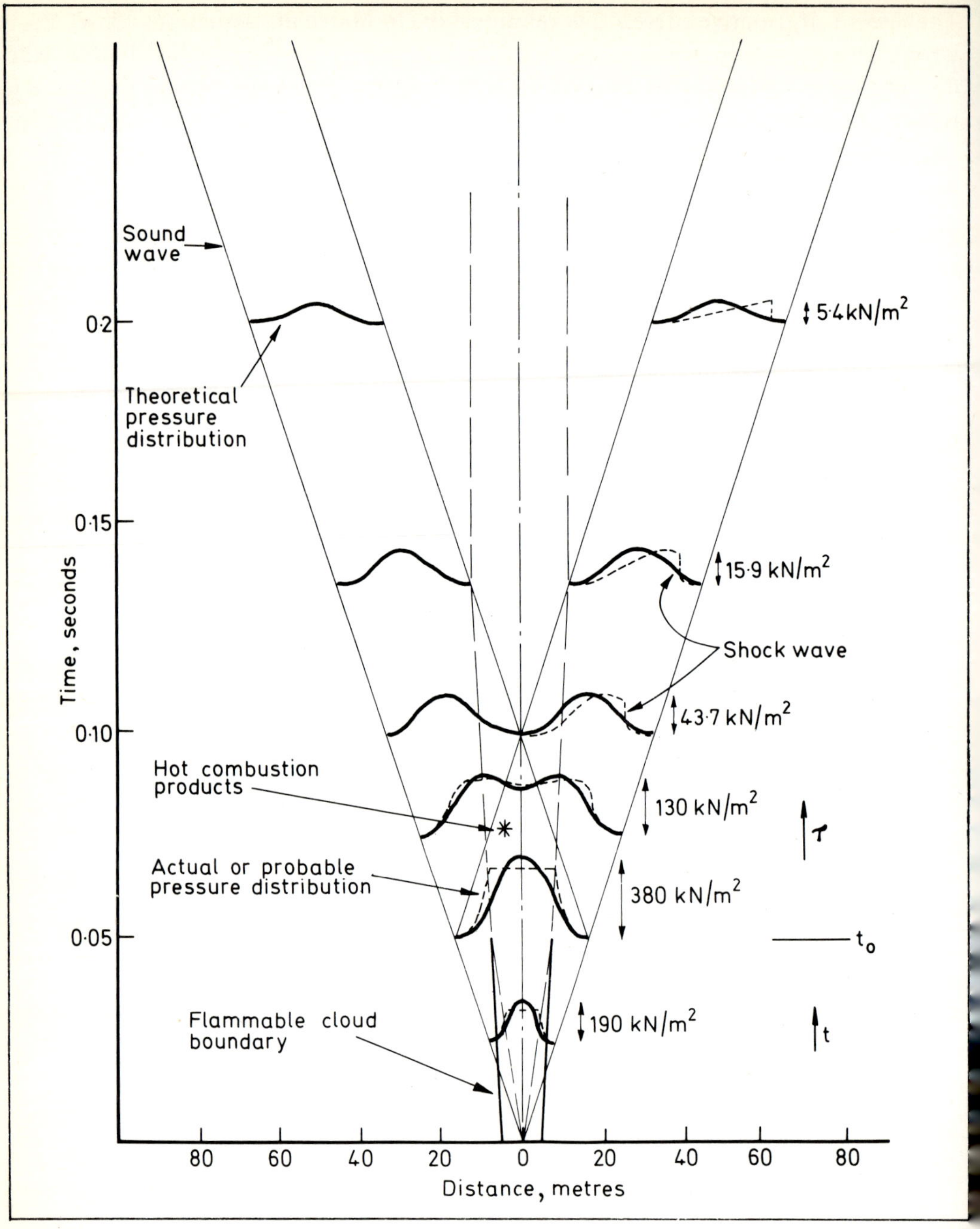

Figure 12 – *Diagrammatic representation of the modes of pressure generation and subsequent dissipation considered in this report. The shapes of the pressure wave profiles are "companions to the cycloid" and the accompanying numbers illustrated are the examples which have been calculated using equations (26) to (30) inclusive of the Appendix. Energy release in this example occupies the time to t_0 and subsequently energy transport and dissipation occurs. The boundaries within which the pressure profiles are constrained are those positions in distance and time related to the sonic transport of pressure information*

The pressure gradient derived from equation (21) is

$$\frac{dp}{dr} = -\frac{\pi \dot{p}_c}{2a}\sin\left(\frac{\pi r}{at}\right) \quad \text{(22)}$$

which has a maximum value of

$$-\frac{\pi \dot{p}_c}{2a} \quad \text{when} \quad r = \frac{at}{\pi}$$

Inserting the value for $\left(\frac{dp}{dr}\right)_{max}$ in equation (20) gives

$$t_s \cong 1.5 \times 10^5/\dot{p}_c \text{ seconds} \quad \text{(23)}$$

The energy contained in the compressed gases is given by

$$E = \int V\,dp$$

in which the relationship between V and p is isentropic. Thus:

$$\int V\,dp = -\gamma \int p\,dV$$

Therefore at any point in time:

$$E = \gamma \int_o^{at} 2\pi r^2 \frac{\dot{p}_c t}{2}\left[1 + \cos\left(\frac{\pi r}{at}\right)\right] dv \quad \text{(24)}$$

for a hemispherical system. Rearranging and substituting,

$$E = \pi\gamma\dot{p}_c t\left(\frac{at}{\pi}\right)^3 \int_o^{\pi} \theta^2 \left[1 + \cos\theta\right] d\theta$$

or $$E = \gamma\dot{p}_c t(at)^3\left[\frac{\pi}{3} - \frac{2}{\pi}\right] = 0.57\dot{p}_c t(at)^3 \quad \text{(25)}$$

or at the conclusion of energy release, $t = t_o$ when $E = E_o$,

$$\dot{p}_c t_o = \frac{E_o}{0.57(at_o)^3} \quad \text{(26)}$$

By way of example, case No 42 may be cited. Assuming that in that case 10 kg of hydrogen in a stoichiometric hydrogen/air mixture were burnt in 0.15 seconds (t_o)—in accord with the *data*[79]—then $E = 1.2 \times 10^9$ joule and $a.t_o = 49.5$ m; therefore $\dot{p}_c \cdot t_o = 1.7 \times 10^4$ Pa (2.5 psig) and $\dot{p}_c = 1.2 \times 10^5$ Pa s^{-1} (16.7 psi/s).

In its subsequent motion the pressure wave has been assumed to travel radially outwards with all points moving at sonic velocity. The wave is assumed to maintain a self-similar profile, of diminishing amplitude but constant pulse duration. The values of pressure exhibited by the wave at any point in time ($t_o + \tau$), after the energy release phase is complete (at t_o), are evaluated assuming conservation of energy within the wave.

The expression defining the pressure arising in the atmoshpere τ seconds after an energy release occupying t_o seconds is:

$$p = \frac{\phi}{2}\left\{\left[1 + \tfrac{1}{2}\cos\left[\frac{\pi(r - a\tau)}{at_o}\right] + \tfrac{1}{2}\cos\left[\frac{\pi(r + a\tau)}{at_o}\right]\right] + \left[\tfrac{1}{2} + \tfrac{1}{2}\cos\left[\frac{\pi(r - a\tau)}{at_o}\right]\right]\right\} \quad \ldots\ldots\ldots\ldots (27)$$

The first square bracketed term has relevance only for values of r lying between zero and $a(t_o - \tau)$ when $\tau < t_o$. The second square bracketed term is valid only for values of r between $a(\tau - t_o)$ and $a(\tau + t_o)$ and when $\tau > t_o$.

As before, the energy content of the pressure wave can be discovered by integration. Two expressions result, one in the time span $o < \tau < t_o$.

$$E_o = \pi\gamma\phi\left(\frac{at_o}{\pi}\right)^3\left[\frac{\pi^3}{3}\left(1 + 3\left(\frac{\tau}{t_o}\right)\right)^2 + 2\pi\left(1 - \frac{\tau}{t_o}\right)\left[\tfrac{1}{2}\cos\frac{2\pi\tau}{t_o} - \cos\frac{\pi\tau}{t_o}\right]\right.$$
$$\left. + \left[\pi^2\left(1 - \frac{\tau}{t_o}\right)^2 - 2\right]\left[\sin\frac{\pi\tau}{t_o} - \tfrac{1}{2}\sin\frac{2\pi\tau}{t_o}\right] - \pi\left(1 + \frac{\tau}{t_o}\right)\right] \quad \ldots\ldots (28)$$

The other applies when $\tau > t_o$

$$E_o = \pi\gamma\phi\left(\frac{at_o}{\pi}\right)^3\left[\frac{\pi^3}{3} - 2\pi + \pi^3\left(\frac{\tau}{t_o}\right)^2\right] \quad \ldots\ldots\ldots\ldots (29)$$

or $$\phi = \frac{E_o}{(at)^3\left[0.57 + 4.4\left(\frac{\tau}{t_o}\right)^2\right]} \quad \ldots\ldots\ldots\ldots (30)$$

Equations (28) and (29) become identical to equation (25) when $\tau = 0$. Similarly, equations (28) and (29) become identical when $\tau = t_o$. Equations (28) and (29) are represented graphically in Figure 8. Figure 12, in addition to indicating the development of pressure, also shows the form of energy dissipation hypothesised. Blast waves transform into shock waves gradually. Those parts of the wave exhibiting the steepest pressure gradient transform first, to be joined progressively by those parts originally of less pressure gradient. ϕ is the maximum instantaneous pressure in the blast wave. The pressure generally in the blast wave, for $\tau > t_o$, is given by:

$$p = \frac{E_o\left[1 + \cos\left[\frac{\pi(r - a\tau)}{at_o}\right]\right]}{(at)^3\left[0.57 + 4.4\left(\frac{\tau}{t_o}\right)^2\right]} \quad \ldots\ldots\ldots\ldots (31)$$

from which:

$$\frac{dp}{dr} = -\frac{\pi E_o \sin\left(\dfrac{\pi(r - a\tau)}{at_o}\right)}{2(at)^4\left[0.57 + 4.4\left(\dfrac{\tau}{t_o}\right)^2\right]} \quad \ldots\ldots\ldots\ldots \quad (32)$$

Similarly, during the energy liberation phase:

$$\frac{dp}{dr} = -\frac{\pi E \sin\left(\dfrac{\pi r}{at}\right)}{1.14(at)^4} \quad \ldots\ldots\ldots\ldots \quad (33)$$

from equations (22) and (26).

These expressions, when substituted into the general form of equation (20), reveal that t_s is a function of r and τ (as well as t, t_o and $\dot{p}_c$). What this indicates is that, with the assumed wave form, there will always be a part of the original blast wave which escapes amalgamation into the shock wave.

The transformation phase is difficult to represent mathematically and I have not attempted to do so. I have, however, considered at the other extreme the shock wave adopting representations of a similar form to those used by others[41, 189]. These give the pressure distribution behind the shock in exponential form, that is:

$$p = p_s\left[1 - \frac{r - a\tau}{r_b - a\tau}\right] \exp\left[-k\left[\frac{r - a\tau}{r_b - a\tau}\right]\right] \quad \ldots\ldots\ldots\ldots \quad (34)$$

Applying energy conservation, as before, then:

$$p = \frac{kE_o\left[1 - \dfrac{r - a\tau}{r_b - a\tau}\right]\exp\left[-k\left(\dfrac{r - a\tau}{r_b - a\tau}\right)\right]}{2\pi\gamma(at)^3\left\{\dfrac{\left(\dfrac{\rho_b}{k} - \dfrac{6}{k}\right)\left(\rho_b - 1\right) - \left(1 - \dfrac{3}{k}\right)\left(\rho_b - 1\right)^2 + \left(\rho_b - 1\right)^3}{(\rho_b - 1)^3} + \left[\dfrac{\rho_b}{k^2} - \dfrac{6}{k^2}\right]\dfrac{e^{-k(1-\rho_b)} - 1}{(\rho_b - 1)^2}\right\}} \quad \ldots \quad (35)$$

where $\rho_b = \dfrac{r_b}{r_b - a\tau}$

and where r_b is the radius at which overpressures vanish.

Baker[41] tabulates values of

$$k, \quad \frac{p_s}{p_o}, \quad \frac{(a\tau)\rho_o^{1/3}}{E_o^{1/3}}, \quad \frac{(r_b - a\tau)\rho_o^{1/3}}{E_o^{1/3}}$$

and other parameters. Correlation with experience indicates that k is reasonably constant (within a factor of 2) in the range

$$0.6 < \frac{(a\tau).p_o^{1/3}}{E_o^{1/3}} < 1000$$

Fuel/air explosives are known to give overpressure durations about twice those for conventional solid explosives such as TNT[190], for the same explosive energy and peak overpressure. $r_b - a\tau$ for shock waves arising as the result of UVCEs must be modified accordingly.

One aspect which becomes immediately obvious on comparing equations of the blast type, equation (31), with their shock wave counterparts, equation (35), is that the amplitude of the shock wave pressure, for identical distances and energies, is larger than the peak blast wave pressure by one or two orders of magnitude. Yet it is the case that blast waves degenerate into shock waves. Is there a region of increasing pressure, or at least sustained pressure, during this process instead of the usual dissipation of amplitude with distance? It is difficult to see how there could be, as gross inertial effects are absent. The conclusion one is forced to is that any untypical wave form results. Untypical, that is, of an explosive shock and untypical in the sense that probably the duration of overpressure is greatly extended without overpressure being augmented. A factor of two is mentioned in the literature for fuel/air explosive devices, but this is probably an under-estimate for UVCEs.

Here, then, we see the principal difference between UVCE blast waves and explosive shock waves. The energy in blast waves is relatively low grade—in pressure terms—but, because of the long durations of the pressure pulses, damage often results which is not dissimilar to that produced by conventional explosive shock waves of much shorter durations and higher pressures. Recognition of these differences enables some reconciliation of TNT equivalents. Before quantifying this matter, however, consideration has to be given to the nature of impulses of different types and their effects. Fortunately, the effects of shock wave type impulses have been considered elsewhere[86] and "isoresponse" curves generated. These are curves of constant response to impulses of different magnitude and duration. Unfortunately, however, they do not allow correlation of impulses of substantially different character. The effect of a wave of the form represented by equations (21) or (31) is to submerge objects in a field of more or less uniformly rising pressure. This creates pressure gradients across these objects which have maximum values in a radial direction from the centre of explosion. The pressures at two radii R_1 and R_2 (where $\tau > t_o$) are given by:

$$p_1 = \frac{\phi}{2}\left[1 + \cos\left[\frac{\pi(R_1 - a\tau)}{at_o}\right]\right] \qquad (36)$$

and

$$p_2 = \frac{\phi}{2}\left[1 + \cos\left[\frac{\pi(R_2 - a\tau)}{at_o}\right]\right] \qquad (37)$$

The nett pressure force felt by an object occupying the space between R_1 and R_2 is thus:

$$p_1 - p_2 = \phi \sin\left[\frac{\pi}{at_o}(R_m - a\tau)\right] \sin\left[\frac{\pi L}{2at_o}\right] \qquad (38)$$

where R_m is the mean distance of the object from the explosion centre and L is its thickness or length in a radial direction from the centre. The pressure force acting is zero when $R_m = a\tau$ (as well as when $a\tau - at_o < R_m < a\tau + at_o$), *ie*, where the crest of the pressure wave spans the object (and outside the pressure wave), and when $L = 0$. The total impulse, I, such a wave provides is not

$$\int_{\tau - t_o}^{\tau} (p_1 - p_2)\, d\tau$$

however, because ϕ changes as the wave moves. Nevertheless, if an appropriate mean value of ϕ is adopted

$$I = \frac{2t_o\bar{\phi}}{\pi} \sin\left(\frac{\pi L}{at_o}\right) \qquad (39)$$

Similarly, the maximum pressure force may be obtained using:

$$(p_1 - p_2)_{max} = \bar{\phi} \sin\left(\frac{\pi L}{at_o}\right) \qquad (40)$$

These impulses and forces are "positive" and are due to the action of the "leading edge" of the pressure wave. Reverse action impulses and forces are experienced following the passage of the pressure peak. These are slightly lower than those causing "positive" phase damage, but may lead, for example, to greater displacements, if damage done during the positive phase allows this. The durations of the positive and "negative" phases are similar also, each being equal to t_o—on the assumptions made herein. t_o is a long period by explosive impulse duration standards—of the order of 0.1 to 1 second. For all practical purposes, therefore, static loading occurs.

The "static loading régime"[86] appropriate to shock waves produces impulsive forces twice the size of those which arise during gradually applied pressure loading. Thus, when a blast wave transforms into a shock wave, the latter is capable of producing greater impulsive pressure loading for a short space until it has dissipated sufficiently.

Beyond the point of transformation, given by equation (23) in temporal terms and restated here in distance terms:

$$R_s\,(= at_s) = 4.95 \times 10^7/\dot{p}_c \text{ m} \qquad (41)$$

no distinction is possible between UVCEs and explosives on the basis of shock overpressure alone. Between the centre of explosion and that point, however, comparison reveals interesting features. Very approximately for explosive shocks:

$$p_s = \frac{250E_o^{1/3}}{R} \text{ Pa} \qquad (42)$$

(This is probably true only for $p_s < 100$ Pa, but it provides a simple basis for comparison and will be used here. For small values of R or large values of E_o or both, p_s can exceed the value predicted by this equation by 100 fold. The deviation is gradual from zero at a value of $E_o^{1/3}/R$ of 0.47 to 100 fold at values of $E_o^{1/3}/R$ of 4.7×10^3.)

Nevertheless, with the obvious restriction which must be placed upon the utility of this expression, it does illustrate why UVCEs are apparently "inefficient" in terms of TNT. Equation (40) gives the maximum positive impulsive pressure associated with UVCE blast:

$$(p_1 - p_2)_{max} = \bar{\phi} \sin\left(\frac{\pi L}{a t_o}\right)$$

Putting $\bar{\phi} = \phi$ for the purpose of this exercise

$$(p_1 - p_2)_{max} = \frac{E_o \sin\left(\frac{\pi L}{a t_o}\right)}{(at)^3 \left[0.57 + 4.4\left(\frac{\tau}{t_o}\right)^2\right]} \quad \ldots \quad (43)$$

$$\text{or} \quad (p_1 - p_2)_{max} = \frac{E_o \sin\left(\frac{\pi L}{R_o}\right)}{R_o^3 \left[0.57 + 4.4\left(\frac{R}{R_o}\right)^2\right]} \quad \ldots \quad (44)$$

where $R_o = at_o$ the radius of the volume affected by the blast wave at the conclusion of the (rapid) energy release phase. Approximately

$$(p_1 - p_2)_{max} = \frac{E_o \sin \frac{\pi L}{R_o}}{4.4\, R_o R^2} \quad \ldots \quad (45)$$

$(p_1 - p_2)_{max}$ is itself a maximum only when $L = \frac{1}{2} R_o \left(\text{or } \frac{at_o}{2}\right)$.

The size of coherent object thus defined is vast in most cases, therefore $(p_1 - p_2)_{max}$ is usually a fraction of its maximum value.

Now the point has already been made that shock waves of long duration produce twice the loading that gradually applied pressures do. Thus, if damage of a given form occurs, it is due either to $\frac{1}{2}p_s$ or $(p_1 - p_2)_{max}$. It now becomes necessary to distinguish the type of energy source: explosive–E_e or UVCE–E_u. Further, the UVCE "efficiency" (η) is defined as

$$\eta E_u = E_e \quad \ldots \quad (46)$$

Again, if m is the equivalent mass of "vapour" involved in the UVCE at the stoichiometric ratio–ϵ, and M_v is the molecular weight of the vapour, then the radius of a hemispherical cloud of stoichiometric mixture (at 0°C and one atmosphere) is:

$$½\left[\frac{m}{M_v}\frac{22.4}{\epsilon}\frac{12}{\pi}\right]^{1/3}$$

Such a cloud would be consumed in time t_o by a flame propagating from a central source at an average speed $\bar{v}$ if

$$t_o = \frac{1}{2\bar{v}}\left[\frac{m}{M_v}\frac{22.4}{\epsilon}\frac{12}{\pi}\right]^{1/3} \qquad (47)$$

Remembering that $R_o = at_o$ then

$$R_o = \frac{a}{2\bar{v}}\left[\frac{m}{M_v}\frac{22.4}{\epsilon}\frac{12}{\pi}\right]^{1/3}$$

or

$$R_o = \frac{727}{\bar{v}}\left[\frac{m}{M_v\epsilon}\right]^{1/3} \qquad (48)$$

Substituting into equation (45)

$$(p_1 - p_2)_{max} = \frac{3.13 \times 10^{-4} E_u \bar{v} \left(M_v\epsilon\right)^{1/3} \sin\frac{\pi L}{R_o}}{m^{1/3}R^2} \qquad \ldots$$

But the UVCE energy release $E_u = m\Delta H$ where ΔH is the heat of combustion.

$$(p_1 - p_2)_{max} = \frac{3.13 \times 10^{-4} E_u^{2/3} \bar{v} \left[\Delta H M_v\epsilon\right]^{1/3} \sin\frac{\pi L}{R_o}}{R^2} \qquad (49)$$

Equating $½p_s$ with $(P_1 - P_2)_{max}$ gives

$$\frac{125\ E_e^{1/3}}{R} = \frac{3.13 \times 10^{-4} E_u^{2/3} \bar{v} \left[\Delta H M_v\epsilon\right]^{1/3} \sin\frac{\pi L}{R_o}}{R^2}$$

ie $$(\eta E_u)^{1/3} = \frac{2.5 \times 10^{-6} E_u^{2/3} \bar{v} \left[\Delta H\ M_v\epsilon\right]^{1/3} \sin\frac{\pi L}{R_o}}{R}$$

alternatively

$$\eta \ngtr \frac{1.56 \times 10^{-15} m\, \bar{v}^3\, \Delta H^2\, M_v\epsilon}{R^3}\ \% \qquad (50)$$

For all its inaccuracies and inadequacies the above does show that comparison of UVCEs and TNT, or explosives in general, is akin to comparing chalk with cheese.

In the near field, within R_s, it indicates that "efficiency" is not only mass dependent, but is much more dependent on the nature of the material ($\bar{\nu}^3 \ \Delta H^2 \ M_v \epsilon$). Equation (50) also suggests that "efficiency" actually falls off rapidly between R_o and R_s.

For values of $\frac{E_o^{1/3}}{R} > 0.47$, p_s may be *very roughly* approximated by

$$p_s = \frac{378.6E^{1/2}}{R^{3/2}} \quad \ldots\ldots\ldots\ldots\ldots \quad (51)$$

This expression, whilst by no means exact, is more accurate than equation (42) over the range

$$0.47 < \frac{E^{1/3}}{R} < 4700.$$

The efficiency obtained using this starting point is:

$$\eta \not> \frac{1.94 \times 10^{-10} m^{1/3} \bar{\nu}^2 \ \Delta H \left[M_v \epsilon\right]^{2/3}}{R} \% \quad \ldots\ldots\ldots\ldots \quad (52)$$

Clearly, overpressure assessment can only be done accurately when damage is unequivocal. For most objects the signs are only unequivocal in the overpressure range 690 to 69 000 Pa (0.1 to 10 psi). Assessment, therefore, is nearly always confined to a relatively narrow pressure band, which exists transitorily over an equivalently narrow stretch of terrain located at a distance from the centre of blast which is directly related to the quantity involved. The application of either of the above expressions for efficiency is not recommended; they merely serve to illustrate the much more complicated form of UVCE yield, expressed as TNT, than has heretofore been presented.

Many more criticisms may be made of the above than have been stated—for example, no mention has been made of the impulsive damage régime of explosives where the pulse duration is so short that the assumption basic to the above is invalid. No doubt, sufficient study will in due course enable an expression for η to be derived with validity in the short pulse régime of conventional explosives.

The Mode of Energy Release

In the previous sub-section a form of pressure wave was *assumed* having characteristics *a priori*, of a kind that experience with UVCEs has suggested is right. The form of pressure wave chosen was one of the simplest (but clearly not the only form) that could be posited. No attempt was made to link the rate of energy release with the developing wave form although the two are clearly inter-related. It is, perhaps, to be inferred that the rate of energy release was constant, as the profile of the wave remains self-similar. In fact, however, constancy of profile implies an accelerating flame.

Equation (25) states:

$$E = 0.57 \dot{p}_c \, a^3 \, t^4$$

If flame speed increases according to

$$v = Kt^{1/3} \qquad (53)$$

then the energy released,

$$E = \int_o^t 2\pi r^2 \, dv \, \Delta h$$

may be expressed as $$E = \frac{9\pi K^3 \Delta h}{8} \int_o^t t^3 \, dt \qquad (54)$$

$$\left(\text{since } v = \frac{dr}{dt} = Kt^{1/3}\right)$$

$$\text{or } E = \frac{9\pi K^3 \Delta h}{32} t^4 \qquad (55)$$

from which, together with equation (25):

$$\dot{p}_c = 4.3 \times 10^{-8} K^3 \Delta h \qquad (56)$$

The flame acceleration is given by $\frac{K}{3} t^{-2/3}$

It is perhaps interesting, although it was not an explicit assumption, that the "arbitrarily" adopted initial blast wave profile should necessarily involve an accelerating flame; it is, after all, generally accepted that flame acceleration is an essential feature of UVCEs. The behaviour of the flame in practice must, undeniably, be complex. It will certainly not choose to obey an arbitrary rule, such as equation (53). Nevertheless, for consistency these considerations could and should be taken further. Equation (47) uses the conception of a mean flame speed $\bar{v}$. It is useful to know what $\bar{v}$ is as a function of K. t_o is the total length of the energy release phase and t any time within that period, therefore

$$\bar{v} = \frac{1}{t_o} \int_o^{t_o} Kt^{1/3} \, dt$$

$$\text{or} \quad \bar{v} = \frac{3K}{4} t_o^{1/3} \qquad (57)$$

alternatively, $$\bar{v} = \frac{3}{4} v_t \qquad (58)$$

Whereas the context of the last sub-section was clearly that of the UVCE, no specific assumption was made there concerning the type of energy release. The mechanics and chemistry of the process were not relevant. It is necessary now to consider these in more detail.

From equations (26) and (30) it is obvious that the rate of energy release has a profound effect on the course of subsequent events. This is another way of stating that flame speed is important. The equations go further than that, however. They enable predictions, when there is no coherent or identifiable flame, such as when spontaneous ignition occurs simultaneously throughout the volume. t is then wholly chemically determined and is short but finite. The speed of sound is also involved and, within or in the near vicinity of a flammable emission, may not exhibit its usual value. Highly significant work in this context was cited earlier. Very dense fogs, because of the change in p *vs* V characteristics they create in the supporting compressible medium, can reduce the velocity of sound by as much as 20 fold[147]. Such extreme effects are not likely in practice, but a 20% reduction in a would lead to a factor of 2 augmentation in blast pressure, equation (26).

Up to this point in the Appendix we have considered an ideal system: hemispherical symmetry and uniform behaviour. Neither assumption is valid in application to real events. Only in the case of catastrophic total failure of a vessel can hemispherical symmetry really be appropriate, but as is obvious from tests[143] behaviour is far from uniform in respect of either dispersion or combustion. The latter is true because the dispersed sources of ignition could not achieve or in any way approximate to central ignition.

Sadee, Samuels and O'Brien[124] have made the point that it is *not* usually the case that ignition occurs within the cloud. If a source of ignition exists, a flammable cloud will ignite on encountering it. It will, under "ordinary" circumstances, always ignite at the edge. It has been suggested that even under "extraordinary" circumstances it will tend to ignite effectively at the edge[21]. Only four cases gave blast (without the aid of building confinement), where three-dimensional release possibly occurred: Nos 1, 14, 38 and 62. Cases 1 and 38 were indeed three dimensional – although the latter was a special case. Not enough is really known about cases 14 and 62 to be sure. The remaining UVCEs all involved clouds having one principal axis or plane (Table 2). Further, the clouds comprised mixtures with air which were far from uniform or uniformly varying. Within the flammable portions of the cloud would have existed regions incapable, *in the ordinary course of events,* of supporting a flame.

Added to this is the effect of variable velocity of sound, which has fundamental significance on pressures developed whilst the blast wave still lies within the flammable cloud. Of course, an additional factor is that the velocity of sound in gases is inversely proportional to the square root of the molecular weight of the medium concerned. This alone can lead to severe turbulence developing in adjacent regions, where pressures following the passage of the sound wave become unequal.

All of this adds to the difficulties in understanding what happened in any particular case. Whatever these difficulties are, however, they are small compared with the problem of predicting UVCE behaviour. One or two general guidelines may be considered.

Since the pressure-affected region, resulting from the consumption of an element of flammable cloud, is centred within that element, a symmetrical blast wave can only result by the subsequent combustion of adjacent elements in such a way that the centre of symmetry does not move. In practice this is possible only with a radially propagating flame from a central source. As this is not a real possibility, the real system has to be defined and the effects of its combustion studied. If the primary (in the sense of first) element of cloud to be ignited is an edge element, then the centre of the pressure-affected region moves in the direction of the flame and the pressure gradient which develops in that direction will be augmented by some power of

$$\frac{a+v}{a-v}.$$

Pressure gradients in the reverse direction will be reduced and those normal to the path of the flame will be unchanged – except that they will exert pressure forces on any fixed object, which will rapidly change in direction.

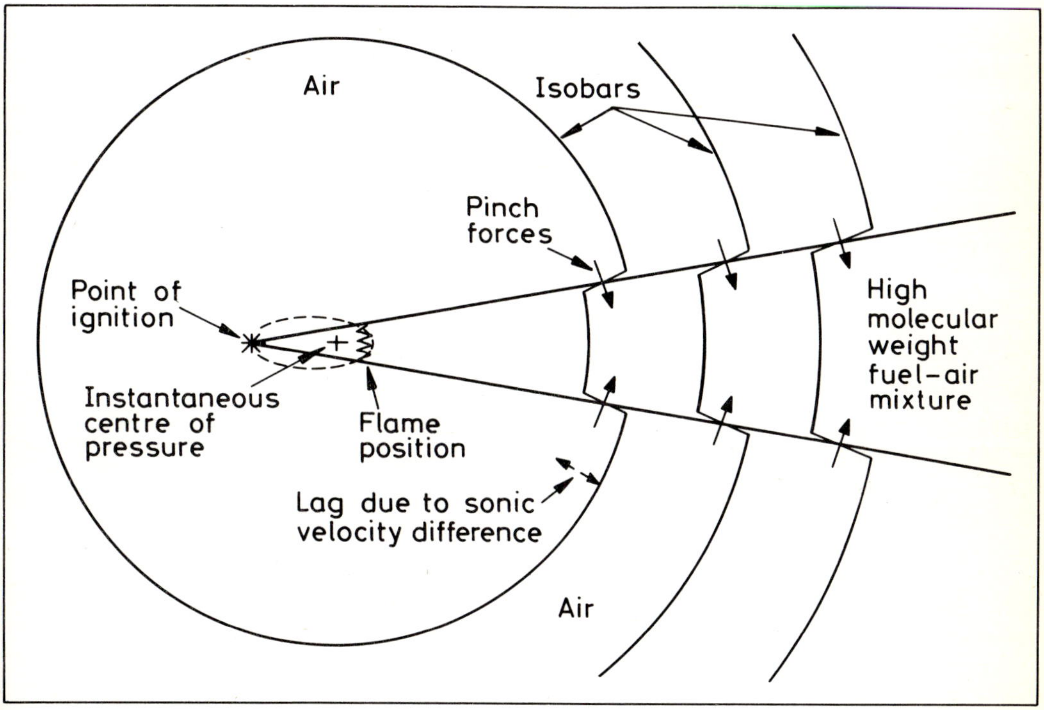

Figure 13 – *This is a diagrammatic illustration of the aerodynamic acceleration effect arising because of the different velocities of sound outside and within fuel/air mixtures – the so-called "pinch" effect referred to in the text. The situation illustrated is one in which the density of the flammable mixture is greater than that of the surrounding atmosphere. The arrows indicating the direction of the pressure force would be reversed, as indeed would be the relative positions of the isobars within and outside the flammable zone, if the flammable mixture had a density less than that of air*

Clearly, flame propagating into a combustible mixture having a predominantly uniaxial or planar distribution will have more influence on the movement of the centre of pressure of the blast wave than a flame propagating from the edge into a three-dimensional cloud. Also there is the effect of those parts of the blast wave travelling in air which are ahead (for high-molecular-weight materials) of the corresponding parts of the blast wave in the cloud, the progressive effect of which would be to "pinch" the as yet unconsumed cloud, in turn producing effects which must be highly relevant. Fortunately, this assertion, *which is universally true,* can be supported by analysis. If a system, Figure 13, is considered which exhibits full spherical symmetry, to the centre of which penetrates a conical sector of circular cross-section and straight axis, the apex of which lies at the centre and if that conical sector comprises a flammable mixture with air and it is ignited at the centre, the blast wave profiles, equation 21, within and outside the sector respectively are given by:

$$p_w = \frac{\dot{p}_c t}{2}\left[1 + \cos\left(\frac{\pi r}{a' t}\right)\right] \quad \ldots\ldots\ldots\ldots \quad (59)$$

$$p_o = \frac{\dot{p}_c t}{2}\left[1 + \cos\left(\frac{\pi r}{a t}\right)\right] \quad \ldots\ldots\ldots\ldots \quad (60)$$

where a' and a are the respective velocities of sound in the two media. At r and at the same instant of time t

$$\Delta P = p_o - p_w = \frac{\dot{p}_c t}{2}\left[\cos\left(\frac{\pi r}{at}\right) - \cos\left(\frac{\pi r}{a' t}\right)\right] \quad \ldots\ldots\ldots \quad (61)$$

where ΔP is the (inward acting) pressure difference exerted across the boundary between the sector and the surrounding atmosphere. This pressure difference acts everywhere at right angles to the "surface" of the conical sector. Equation (61) can be simplified to

$$\Delta P = \dot{p}_c\, t \sin\left[\frac{\pi r}{2at}\left(\frac{a}{a'} + 1\right)\right] \sin\left[\frac{\pi r}{2at}\left(\frac{a}{a'} - 1\right)\right] \quad \ldots\ldots \quad (62)$$

$$\text{since} \quad \frac{a}{a'} = \left[\frac{T\gamma \; M'}{M \; T'\gamma'}\right]^{1/2} \quad \ldots\ldots\ldots\ldots \quad (63)$$

and is always close to unity; equation (62) can be further simplified:

$$\Delta P = \frac{\dot{p}_c \,\pi\, r}{2a}\left\{\left[\frac{T\gamma M'}{T'\gamma' M}\right]^{1/2} - 1\right\} \sin\left\{\frac{\pi r}{2at}\left(\left[\frac{T\gamma M'}{T'\gamma' M}\right]^{1/2} + 1\right)\right\} \quad (64)$$

To give some idea of the magnitude of the pressure difference, consider again case No 42. The maximum value of ΔP occurs when

$$r \cong \frac{at}{2.}$$

From equation (64) therefore: $\Delta P \sim -\ 0.12\ \dot{p}_c t$. The negative sign implies an outward acting pressure difference.

Another example is stoichiometric cyclohexane. The stoichiometric assumption – made elsewhere – is not essential, but it is convenient. The molecular weight of a mixture of cyclohexane in stoichiometric proportions in air (M') is – referred to air (M=29) – 30.25. Again using equation (64):

$$\Delta P = 1.68 \times 10^{-2}\ \dot{p}_c t$$

ΔP in this case acting inwards.

Clearly, in both examples the magnitude of the effect created increases both with time and rate of energy release. For materials of lower molecular weight than air ΔP is negative and conversely for those materials of higher molecular weight. In the latter case the effect is exactly analogous to squeezing a toothpaste tube and the result is presumably similar; the gas is driven forward (radially outward from the centre of symmetry) by the "external" pressure.

The acceleration (f) thus induced in a disc-shaped slice of the conical sector at a radius r from the centre of symmetry is:

$$f = \frac{2\ \Delta P \cos\alpha}{r\rho} \qquad \ldots\ldots\ldots\ldots \qquad (65)$$

where α is the semi-angle of the cone sector. From this equation it is clear that the thinner the cone, the greater the rate of acceleration imparted to the combustible mixture. Equation (65) can be improved on by integrating the effect of ΔP over the entire blast-wave-affected zone and dividing the resultant force by the total mass of the conical sector. The result is that at any time t after ignition the average radially outward acceleration (f) experienced by the gases in the cone would be:

$$f = \frac{3\dot{p}_c}{a\rho}\left[1 - \frac{2}{\pi^2}\right]\left\{\left[\frac{T\gamma M'}{T'\gamma' M}\right]^{1/2} - 1\right\}\cos\alpha \qquad \ldots\ldots \qquad (66)$$

For the cyclohexane example presented above:

$$f \ngtr 1.15 \times 10^{-4}\ \dot{p}_c\ m/s^2 \left[\ \dot{p}_c \text{ in Pa s}^{-1}\ \right]$$

It is interesting to note that acceleration effects for high-molecular-weight flammables cease for "cone" semi-angles greater than 90°.

Plainly, the value of $\dot{p}_c$ determines what happens. Evidence at Flixborough suggested that $\dot{p}_c$ substantially exceeded 7 MPa s^{-1}. The vapour cloud acceleration in that case, had there been a well-defined uniaxial cloud, would have been 800 m s^{-2}

It is also interesting to note that for high-molecular-weight flammables this effect "couples" with the driving force behind the event. Low-molecular-weight flammables decouple, possibly to the extent that unidimensional combustion is extinguished (by enhanced mixing and dilution ahead of the flame – case 42 perhaps).

Again we have left out the flame. This time, however, it was justified, because none of the "pinch" effects are directly chemistry or combustion dependent. But the reverse is not true. As set out in the section on combustion fundamentals, there are two aspects to flame speed: that residing in the intrinsic character of the flame system (burning velocity, temperature, pressure, composition and turbulence) and that imposed upon the system (translational movement). In what has been set out above lies the whole answer to the development of high flame speeds – they are overridingly translational, not, as has been supposed hitherto, the result of inherent factors.

The method by which high flame speed is attained is irrelevant to stationary observers facing it. Patently, this system of combustion dynamics is unsymmetrical. It is also non-steady-state. Because of the confinement imposed on the flammable mixture, it may adopt some of the characteristics of combustion in a tube. If it does it may detonate.

A conical distribution of flammables is the norm for a forceful discharge, but it would be expected that a relatively thin plane of combustible mixture might behave similarly if it were ignited at the edge. What seems clear is that in respect of this particular mechanism "edge" ignition of a hemi-spherical or spherical cloud could not cause flame acceleration. This supplements the knowledge that to the present flame acceleration does not seem to occur in hemi-spherical or spherical symmetries with central ignition.

In the more conventional theoretical approach mechanisms of flame acceleration have been suggested which skirt the fundamental issue of precisely how this happens, since laboratory work on flame speed enhancement by turbulence in three-dimensional or "unconfined" systems does not produce anything like the necessary final flame speed (2 orders of magnitude increase over "normal" speed). The leading theoretical workers in this field take *as their starting point* flames already accelerated to high speed[128,191,192]. It may, of course, be right to hurdle the fundamental difficulty in search of the solution to broader issues. Nevertheless, the issue remains and, whilst it remains, one is permitted to ask: is the model right and is there a mechanistic difficulty, or is the model wrong and, if so, what are the mechanics?

What is the model? Kuhl *et al*[128] is the basic treatise, although this is itself based on a work of far greater depth[193]. It extends the classical work of Taylor[14] and claims to overcome the dilemma first enunciated by Taylor of the incompatibility of the "blast wave" and acoustic solutions at the transition of the former to the latter. The reason for this difficulty is explained: flames, because of their relatively low propagation velocities, gave solutions between the "blast wave" and acoustic fields. The model retains Taylor's spherical piston conception, wherein pressures generated by movement of the piston themselves generate a shock wave whose characteristics are wholly governed by the piston speed. The spherical piston is, of course, the expanding envelope of hot products of combustion. At 40 m s^{-1} flame speed the pressure just ahead of the flame front is calculated to be 176 kPa (25 psig) and at the shock front 138 kPa (20 psig). These figures are taken directly from

Figure 11 of Kuhl *et al*[128] although Clancy[194] quotes a somewhat lower value of pressure taken from "their relationships". Geiger[191] presents a pictorial representation (Figure 9 of the reference) of pressure and velocity in a burning unconfined combustible mixture which seems inconsistent with the results given in Kuhl *et al*[128], although Geiger would indeed seem to have worked from Kuhl. For example, he shows, in spherical co-ordinates, a system in which the flame is propagating at 80.5 m s^{-1} and developing only 9 kPa (1.3 psig) at the flame, which manifests a continuous fall in pressure to zero at the pressure wave front (the *at* boundary). No shock wave is indicated. Geiger's system in this respect is similar to that proposed earlier herein. In the central region, however – within the flame envelope – pressures are, in his system, uniform. Geiger, in his Figure 9[191], does show a small pressure discontinuity at the flame. This pictorially represents the pressure effect created by the change in momentum through the combustion zone.

One thing is immediately apparent from these works: scale of event is not a material factor in the effect produced. This is not a criticism, however, as flame acceleration would occupy some space and thus the effects would accumulate. But again the point arises, how does thc flame accelerate? Unless this is known or can be predicted, the precise effect of scale cannot be ascertained. This issue is predominant – unless it is resolved, no degree of confidence can be placed in any boundary drawn between what is a "safe" vapour cloud and what size will produce blast.

The "classical" theory of UVCEs involves an ideal combustion system. This does not happen in practice and, where it has been contrived by experiments[137], it has not produced significant pressure effects. Aspects of non-ideality have been considered above. Two "non-classical" factors are the imploding flame and radiation induced spontaneous ignition[21]. Fortunately, both have received attention recently[195] although from the particular standpoint of transition to detonation. The former mechanism is likely to give an accelerating flame in a *laminar* system only for Lewis numbers in excess of unity. As the practical system is not laminar, no direct correlation with the published theory[196] should or could be anticipated. On the other hand, as the real significance of the Lewis No is the ratio of conduction- to diffusion-controlled energy transport and as both would be expected to alter in the same way with increasing turbulence[197] the *pure* Lewis No argument for pressure effects from a spherically converging flame is difficult – if not unconvincing. Nevertheless, as the effective value of Lewis No is dependent on the effective transport of thermal energy, the role of radiation in such a (practical) system cannot be ignored.

Ordinarily in combustion systems combustible gases are not exposed for prolonged periods to flame radiation. In the event, however, of a larger volume of combustible vapour being surrounded by a thick and highly radiant flame envelope relatively prolonged exposure could occur *depending* upon the flame speed. The matter has been considered previously[21,198] and is plausible "although detonations due to a thermal mechanism are fairly unlikely[195]."

Effects of Blast

These vary depending upon the type of event considered. Conventional explosive blast effects are well documented[158,160,163,165,199,200] and may be divided into direct effects and indirect effects (which include missiles and displacement effects). Two categories of objects may also be distinguished: "inanimate" objects – including animals – and human beings. The effects of conventional explosives will be considered first.

Conventional condensed phase explosives

The point is made in the text that the response of any system to a force depends not only on the characteristics of the force but also the characteristics of the system. Of the systems relevant to this study all have inertia and most are restrained; the restraint taking the form of elastic deformation initially, followed by plastic deformation.

As the response of *systems* is a factor common to their behaviour independently of the type of pressure wave form acting upon them, the following is a presentation of the relevant theory. Analysis of the response of a simple unrestrained mass has been performed[158] and attested experimentally. The theory of motion is elementary. According to Bowen *et al* and based on Newton's Second Law of Motion:

$$\dot{v} = \frac{dv}{dt} = \frac{\rho_a \, (u - v)^2}{2} \, \beta \quad \ldots\ldots\ldots\ldots\ldots\ldots\ldots\ldots \quad (67)$$

where u = velocity of air,
v = velocity of object,
$\dot{v}$ = acceleration of object,
ρ_a = density of air,

and β = acceleration coefficient defined as

$$\beta = \frac{SC_d}{m},$$

where S = area of object presented to the wind
C_d = drag coefficient of the object and
m = the mass of the object.

Before equation (67) can be used diagnostically or predictively, it has to be modified according to the properties of the blast wave. As the velocity of the air, u, is not the same as the blast wave velocity, $\dot{x}$, and as the relative velocity of the object to the blast wave determines for how long the object suffers an accelerating force, a factor allowing for this effect has to be included in equation (67). The complete equation of motion is therefore:

$$\frac{dv}{dt} = q\beta \left(\frac{u - v}{u}\right)^2 \left(\frac{\dot{x}}{\dot{x} - v}\right) \quad \ldots\ldots\ldots\ldots\ldots\ldots\ldots\ldots \quad (68)$$

where q is the dynamic (wind) pressure $\left(= \frac{\rho_a u^2}{2}\right)$.

Bowen *et al* transformed this equation into the dimensionless form:

$$\frac{dV}{dZ} = QB \left(\frac{U-V}{U}\right)^2 \left(\frac{\dot{X}}{\dot{X}-V}\right) \quad \ldots\ldots\ldots\ldots\ldots\ldots\ldots \quad (69)$$

by setting $Q = \dfrac{q}{p_o}$,

$$B = \frac{\beta p_o \, t_u^+}{a}$$

$$U = \frac{u}{a},$$

$$V = \frac{v}{a}$$

$$\dot{X} = \frac{\dot{x}}{a}$$

and $Z = \dfrac{t}{t_u^+}$

These authors also used Brode's classical analysis[201] of blast waves to formulate, for computational purposes, the forms of equations for Q and U as functions of Z and

$P_s \left(= \dfrac{p_s}{p_o}\right)$ –

the parameters were, however, empirically determined:

$$Q = Q_s \left(1 - Z\right) \left[F \exp(-\zeta Z) + G \exp(-\delta Z)\right] \ldots\ldots\ldots\ldots\ldots\ldots \quad (70)$$

where $Q_s = \left(\dfrac{2.5\,P_s^{\,2}}{7 + P_s}\right) \left(\dfrac{1 + 2 \times 10^{-8}\,P_s^{\,4}}{1 + 10^{-8}\,P_s^{\,4}}\right)$

$F = 1.186\,P_s^{\,1/3}$ *for* $P_s < 0.6$

or $F = 1.0$ *for* $0.6 \leqslant P_s \leqslant 1.0$

or $F = \left(10^4\,P_s^{-1/4}\right) \Big/ \left(10^4 + P_s^{\,2}\right)$ *for* $P_s > 1.0$

$G = 1 - F$

$\zeta = 0.25 + 3.6\,P_s^{1/2}$

and $\delta = \left(7 + 8\,P_s^{1/2}\right) + 2\,P_s^{\,2} \Big/ \left(240 + P_s\right)$

and $U = U_s \left(1 - Z\right) e^{-\nu Z}$ $\ldots\ldots\ldots\ldots\ldots\ldots\ldots$ (71)

where $U_s = P_s \left(1 + P_s^{1/2}\right)$

and $v = P_s^{1/3} + 0.0032\, P_s^{3/2}$

Bowen *et al* used:

$$\dot{X} = \frac{3}{5} U + \left[1 + \left(\frac{3}{5} U \right)^2 \right]^{1/2} \qquad \text{(72)}$$

in their computations and the graphical relationship between t_u^+ and t_p^+ (*ie* wind duration *vs* overpressure duration) to evaluate Z, t_p^+ being itself derived from

$$\log_{10} t_p^+ = 5.7995 + \frac{1}{3} \log_{10} W - 0.2957 \log_{10} p_s - 0.0376 \log_{10} P_o - \log_{10} a$$

This dimensional equation is in mixed units, using pressures in pounds per square inch, velocity in feet per second, explosive yield in kilotons TNT and deriving t_p^+ in millisecond multiples. This equation converts, in consistent SI units, to:

$$t_p^+ = \frac{36.35\, W^{1/3}}{p_s^{0.2957}\, P_o^{0.0376}\, a} \text{ seconds} \qquad \text{(73)}$$

The graphical relationship referred to above reveals that the wind duration progressively exceeds the overpressure duration as shock pressures increase (up to 3 atmospheres overpressure) to a maximum excess duration of 78%.

Only "wind" forces were considered in the quoted work – a view which appears to have been vindicated by the experimental results the authors obtained. The experiments are described elsewhere in detail[165] but relate to the effects produced by blast waves from nuclear detonations (the Plumbbob tests).

In these tests various "unrestrained" objects were exposed to the blasts and their subsequent motions analysed and correlated with their βs. For most objects β depends upon orientation. This is certainly true of animate objects including man. Calculated values of β for a 76-kg (168-lb) man vary from 1.06×10^{-2} m^2 kg^{-1} for a standing man facing the blast to 1.29×10^{-3} m^2/kg for a prone man aligned with the blast direction.

A complicating factor is that values of β change during the motion due to tumbling. The dummies used in the experiments, although their centres of mass followed parabolic trajectories, executed partial back-summersaults in which they passed from the standing position through the prone position whilst still accelerating. A time-dependent β cannot be conveniently accommodated in the theories so an effective β was attributed to the tumbling man to enable a reasonable correlation with actual behaviour. The effective β ascribed to a man subjected to an explosive derived blast is 6.13×10^{-3} m^2/kg.

The authors[158] cite values of β for various other objects including window glass, but comment that framed window glass in the open air gave different results from framed glass actually installed in buildings. In the latter case the fragments behaved as if the parameters controlling their motion were appropriate to the existence of more than twice the incident pressure at the position of the window frame. Correlations of the movement of glass in houses was found to be satisfactory if the controlling parameter, p_s, was taken to have the reflected magnitude.

In the work cited above the interest was in so-called tertiary effects of blast, *ie*, whole body displacement. The effects of blast on the human body have for some time been categorised as primary, secondary and tertiary by Zuckerman[202] and Glasstone[148]. Primary (or direct) effects of blast will be considered shortly. Secondary (or missile) and tertiary effects are indirectly related to blast in that injuries which result are caused not by the blast pressure wave itself but by objects thrown by the blast wave, or stationary objects into contact with which the human body is thrown.

The second class of behaviour of objects is where restraint exists, coupled with inertia, to control motion subsequent to the arrival of a blast wave. This form of response has been considered by a number of workers[86,149,150,203] who have in turn considered a number of different incident pressure wave forms, but only Mainstone[149,150] to my knowledge has extended the treatment to include "plastic" deformation consequent upon deflection beyond the elastic range of strain. Mainstone does not, however, deal with the sudden-rise pressure profile encountered with explosives. The treatment of Strehlow and Baker[86] considers the motion of a mass, m elastically restrained in such a way that the restraining force appropriate to any deflection is proportional to that deflection, in response to an external force exponentially decaying behind an initial shock:

$$m\ddot{x} + Kx = S\,p_s\,e^{(-t/t_l)} \qquad \ldots\ldots\ldots\ldots \qquad (74)$$

where t_l is the characteristic loading time, S is the effective surface over which P_s acts and K is the constant of elastic proportionality

$$\text{or} \quad x = \left(\frac{Sp_s}{K}\right) \frac{(\omega t_l)^2}{1 + (\omega t_l)^2} \left[\frac{\sin\omega t}{\omega t_l} - \cos\omega t + e^{(-t/t_l)}\right] \qquad \ldots\ldots \qquad (75)$$

in which

$$\omega = \left(\frac{K}{m}\right)^{1/2}$$

From the diagnostic viewpoint interest centres on maximum displacements as a function of the parameters of both the shock wave and the responding system. The asymptotic values are of interest in that they indicate the behaviour under extremes of loading rates. When t_l is less than 0.2,

$$x_{max} = \frac{Sp_s\,t_l\,\omega}{(Km)^{1/2}} \qquad \ldots\ldots\ldots\ldots \qquad (76)$$

and when t_l is greater than 100

$$x_{max} = \frac{2S\,p_s\,\omega}{K} \qquad \ldots\ldots\ldots\ldots\ldots\ldots \qquad (77)$$

Now $Sp_s t_l$ is in fact the exact integral of the imposed force-time expression and is therefore the impulse, I, delivered by the force. Equation (76) therefore becomes:

$$x_{max} = \frac{I\omega}{(Km)^{1/2}} \qquad t_l < 0.2$$

$$\text{or} \quad x_{max} = \frac{I}{m} \qquad \ldots\ldots\ldots\ldots\ldots\ldots \qquad (78)$$

From these two expressions for x_{max} it can be seen that when the loading time, t_l, is long, *ie* quasi-static, the response is proportional to $2p_s$, but when t_l is short, *ie* impulsive, the response is proportional to the impulse. Strehlow and Baker[86] presented a relationship which, it is claimed, will enable fairly accurate correlation with a simple system:

$$(\bar{P} - 1)\quad(\bar{I} - 1) = C \qquad \ldots\ldots\ldots\ldots\ldots\ldots \qquad (79)$$

$$\text{where } \bar{P} = \frac{25p_s}{x_{max}K} = \frac{2}{\left[2 - \exp\left\{-\frac{(\omega t_l)^2}{100}\right\}\right] \tanh\,(\omega t_l)} \qquad \ldots \qquad (80)$$

$$\text{and} \quad \bar{I} = \frac{I}{x_{max}(Km)^{1/2}} = \frac{\omega t_l}{\left[2 - \exp\left\{-\frac{(\omega t_l)^2}{100}\right\}\right] \tanh\,(\omega t_l)} \qquad \ldots \qquad (81)$$

C is a number dependent upon the particular characteristics of the system considered.

Allan and Athens[203] have also studied the shock loading régime but dealt at greater length with a triangular pressure pulse (of isosceles form) and the (elastic) response of structures exposed to it. They predicted that single triangular pulses of durations less than a quarter of the natural period of the structure would produce deflections less than the static deflection under a constant load equivalent to the maximum pressure of the waveform. For pulses of duration greater than one quarter and less than twice the natural period, however, enhanced deflections up to 50% more than would be found with static loading would be encountered. Nevertheless this is still less than the 100% increase in deflection arising from the *quasi*-static load produced by a shock impulse of long duration.

Mainstone's[149,150] principal interest when he produced his analysis of structural response to pressure was in explosions within buildings. These phenomena develop pressure gradually and therefore do not generate the saw-tooth wave form considered by Strehlow and Baker. One of the wave forms Mainstone does treat, however, is the triangular profile also considered by Allan and Athens, although he extends the treatment into the plastic range of behaviour. For the particular case of wholly elastic response Mainstone confirms Allan and Athens' predictions of maximum augmentation of strain of about 50% at a pressure pulse duration of slightly less than the natural frequency of the affected (simple) structure. It is of interest to note, however, that significant subsequent plastic deformation (within design tolerances) completely eliminates this further strain. Naturally, it does not eliminate all the strain, however, and permanent deformation probably results.

Mainstone's pressure wave forms, in addition to the triangular pulse mentioned above, have the profiles which result typically from vented explosions. One of these has a sine-wave shape similar to that developed herein to describe UVCE blasts and will therefore be considered in the next sub-section.

Before considering primary and secondary effects of blast on persons, it is appropriate to mention the explosive blast effects upon – tertiary, and caused by – secondary, minute particles which are inevitably present in the environment. These particles are generally hard and because they are small they achieve very high velocities. The effect of their subsequent impact upon larger objects is to cause characteristic pitting and abrasion, especially close to the focus of the explosion.

Primary effects of blast have been studied extensively. The Lovelace Foundation in the United States is the principal source of unclassified data. Very many papers have been published in the open literature[163]; additionally detailed technical reports are available. One of the latest publications[204] tidies up and logically correlates many of the earlier results.

The primary effects of blast on animals and men are the physiological changes which are induced directly by pressure loading. These are often difficult to differentiate as rarely do significant primary effects occur in isolation from secondary and tertiary effects. Furthermore the correlation of injuries and cause of death are not straightforward, because they are not necessarily the same from individual to individual. Meaningful data is essentially statistical and plainly controlled (*ie* non-warfare) situations of study are not generally available.

Considerable work has been done, however, with animals ranging in size from mice to cattle. This work is statistical and correlations based upon 50% 24-hour survival have been produced for exposure to sharp rising overpressures of various durations. The data indicates two distinct survival patterns based upon the natural sizes of the animal species tested. The division between those two survival patterns occurs at an animal size close to the rabbit and the cat, the rabbit falling into the class of smaller animals. Plainly man would fit into the class of larger animals, thus results obtained with small animals cannot be directly related to man's tolerance to explosive blast.

The 50% mortality in 24 hours criterion is considered to be realistic, because pure blast injuries are generally not chronic in nature and 24 hours, in warfare, is a typical period between injury and effective treatment.

According to White *et al*[205] survival of small animals, on the above criterion, is likely for exposure to peak reflected over-pressures less than about 1.9×10^5 Pa (28 psi), independent of duration, but for durations shorter than 0.5 ms impulses below 35 Pa s can be tolerated. For large animals, including man, the corresponding figures are 3.3×10^5 Pa (48 psi), 3 ms and 520 Pa s. A computed mean correlation of the data which could be applied to a prone 70 kg man at normal atmospheric pressure is:

$$p_r = 4.23 \times 10^5 \left[1 + \frac{4.34 \times 10^{-3}}{t_+^{\ 1.064}} \right] \quad Pa \qquad \ldots\ldots\ldots\ldots\ldots\ldots \qquad (82)$$

where p_r is the peak reflected overpressure *(Pa)* and t_+ is the duration of the positive phase of the pressure wave (seconds).

In as much as body orientation influences translational effects, so does orientation affect the direct pressure effects of blast. The worst situation arises for a man standing against a vertical reflective surface where the peak reflected pressure for the 50% survival criterion would be reduced by about two thirds. A man standing in the open away from reflecting surfaces would expect the same survival criterion at about half the peak reflected pressure which would be associated with 50% survival if he were prone. Very severe injuries are associated with the 50% survival condition. For example, in all cases the peak reflected overpressure which would cause threshold lung damage is about one-fifth of the corresponding 50% survival overpressure.

Hitherto, the main consideration, as was strictly appropriate, in this sub-section has been the "classic" saw-tooth pressure profile. Digressions have occurred in which other waveforms have been considered despite their perhaps more appropriate inclusion in the next subsection. Nevertheless this view would not strictly be correct. Whilst one can arbitrarily divide pressure wave forms into two categories – those with sharp rise and those with gradual rise profiles – the sharp rise profiles do transform into their gradual rise counterparts even close to the focus of blast where strong structures having relatively large volumes but small openings are encountered.

An important distinction must be made before continuing. Man, in common with all animals, can tolerate very high pressures of long duration (seconds) so long as the rates of pressure rise and of pressure decay are not excessive. The tolerances to pressures, indicated above, are where the rates of initial pressure rise is effectively infinite. Enhanced tolerance to pressure has been demonstrated even with fairly rapid pressure rise (achieved in not less than 0.02 s) and high subsequent and sustained pressures. For example[206], dogs have survived 1.7×10^7 Pa s^{-1} followed by 5.2×10^5 Pa sustained for 10 seconds. Rats on the other hand have been exposed to slightly lower rates of pressure rise, 5.7×10^6 Pa s^{-1} but with the pressure taken to higher values – 2.8×10^6 Pa and sustained for up to 40 seconds. The important factors for survival in these tests were rates of fall of pressure and period for which the high pressure had been maintained. The data suggests that exposure to high pressures for periods of a few tenths of a second is survivable so long as the rates of pressure rise and decay do not exceed about 10^7 Pa s^{-1}.

Again the issue is one of *survival*. Injuries will occur and must be expected under extreme conditions. These consist in the main of eardrum rupture, sinus haemorrhage and in some cases lung haemorrhage.

Also important in this general context is the wind effect that can arise in an enclosure when subjected to an external pressure pulse of considerable duration. An initial rate of pressure rise of 10^7 Pa s^{-1} in a room of 27-m^3 volume equipped with openings of aggregate area 5m^2 will generate wind velocities as high as the speed of sound. Such winds will tend to affect those items directly in their path; the moving column of air will be cohesive and persistent. There is little doubt that persons caught in such a wind would be killed, as was demonstrated with animals in the "teapot" experiments[165]. Large temperature variations of short duration would also occur – even in the unlikely event that external temperatures were not high – due to adiabatic compression and decompression within the space, although these would not cause more than discomfort and thus be trivial by comparison.

Unconfined vapour cloud explosions

The effects of UVCEs may be predicted using TNT equivalent data. This technique is widely used and is considered reliable in application outside the burning cloud boundary. No incident is known of, however, where the technique has been used *and* a UVCE has occurred. It is considered inappropriate for use inside the burning cloud because, there, errors would magnify and the view is that enormously stronger structures would be predicted than are in fact necessary. Whether that is a view which is cogent will be examined shortly. At the present time it is known, however, that certain large companies operating in the chemical, petrochemical and refinery fields have tentatively adopted the criterion that structures on the periphery will experience no more than 10^5 Pa for a period of no more than 20 ms should a UVCE occur or, if higher or lower overpressures do arise, the resultant impulse will not exceed 2 kPa s. It is required additionally, however, that the maximum overpressure and impulse values within these constraints should be accommodated by the strength and resilience of the proposed structures.

Another approach is to use the Kuhl, Kamel and Oppenheim[128] method which requires that an initial guess be made of the final flame speed and, possibly more importantly, of the rate of acceleration of the flame. Undoubtedly, such guesses could be made with some confidence based on previous incidents, but no straightforward working engineering-type equations have been published to date, which enable the prediction of factors such as the magnitude and maximum rate of pressure rise or the total duration of excess pressures. It is to be hoped that the book[207] presently in preparation by Dr George Munday will provide what is required.

For the present, therefore, I return to the "theory" outlined in the previous subsections of this Appendix. It is worth reiterating that the sole justification of this approach is that it can be used to predict effects with some accuracy both in areas where other techniques can be applied and in regions where those techniques are not applicable. The theory, whilst being arbitrary, is, nevertheless, consistent – it follows, therefore, that predictions (based upon good data) which are surprising must not be discounted out-of-hand. To do so would be unscientific and could be dangerous.

Equation (26) enables the maximum pressure to be estimated:

$$\dot{p}_c t = \frac{E}{0.57\,(at)^3}$$

from which $\dot{p}_c$ may be determined:

$$\dot{p}_c = \frac{E}{0.57t\,(at)^3}$$

Equation (21) yields on differentiation:

$$\frac{dp}{dt}\ (=\dot{p}) = \frac{\dot{p}_c}{2}\left[1 + \cos\left(\frac{\pi r}{at}\right) - \left(\frac{\pi r}{at}\right)\sin\left(\frac{\pi r}{at}\right)\right] \quad \ldots\ldots\ldots \quad (83)$$

Or, alternatively, the rate of pressure increase ($\dot{p}$) at any distance $r > at_o$ from the centre may be found from:

$$\frac{d\phi}{d\tau}\ (=\dot{p}) = -\frac{5.02\,\dot{p}_c\,\frac{\tau}{t_o}}{\left[0.57 + 4.40\left(\frac{\tau}{t_o}\right)^2\right]^2} \quad \ldots\ldots\ldots\ldots\ldots\ldots \quad (84)$$

What rates of pressure rise occur in practice? These are best assessed by the differential pressure effect on relatively strong but vented containers – photographs 12, 13 and 14. During external pressurisation the flows of gas (assumed to be air) into the container are given by the following expressions:

$$\dot{m} = \frac{9.375 \times 10^{-2}\ A\ P}{\sqrt{T}}\ w^{0.714}\left[1 - w^{0.286}\right]^{1/2} \quad \ldots\ldots\ldots\ldots \quad (85)$$

for values of $w\left(=\frac{P_i}{P}\right)$ greater than 0.528 and

$$\dot{m} = \frac{2.426 \times 10^{-2}\ A\ P}{\sqrt{T}} \quad \ldots\ldots\ldots\ldots\ldots\ldots\ldots\ldots\ldots\ldots\ldots \quad (86)$$

for values of w less than 0.528 and where A is the aggregate area of openings in the skirt.

Concomitantly, pressures within the container will rise due to the increasing mass content. From an initial mass content m_o at an initial pressure P_o the internal pressure at any instant may be represented by

$$P_i = \frac{(m + m_o)\,P_o}{m_o} \quad \ldots\ldots\ldots\ldots\ldots\ldots\ldots\ldots\ldots\ldots\ldots \quad (87)$$

where $m + m_o$ is the mass content at that instant. As we wish to follow the course of the differential pressure ΔP ($=P-P_i$) it is necessary to formulate m. This can be done using equations (85) or (86), but it can be done analytically only with equation (86).

If it is assumed that the external pressure is a linearly increasing function of time (not an essential assumption) then

$$m = \frac{1.213 \times 10^{-2} A \dot{p} \Delta t^2}{\sqrt{T}} \quad \ldots\ldots\ldots\ldots \quad (88)$$

Using equations (87) and (88) (modifying 87 to

$P_i = \dfrac{mP_o}{m_o}$ to "compensate" for assumption of sonic flow):

$$\Delta P = \dot{p}\Delta t - P_o \left[\frac{1.213 \times 10^{-2} A \dot{p} \Delta t^2}{m_o \sqrt{T}} \right] \quad \ldots\ldots\ldots\ldots \quad (89)$$

I[21] have presented a similar relationship, although slightly differently derived and not "compensated" for sonic flow:

$$\Delta P = \left[\dot{p} - \frac{2.434 \times 10^{-2} A P_o}{m_o \sqrt{T}} \left(P_o + \frac{\dot{p}\Delta t}{2} \right) \right] \Delta t \quad \ldots\ldots \quad (90)$$

Both of these equations are quadratic in Δt; accordingly, real solutions for Δt may be obtained only when:

$$\dot{p} > \frac{4.852 \times 10^{-2} A P_o \Delta P}{m_o \sqrt{T}} \quad \ldots\ldots\ldots\ldots \quad (91)$$

in the case of equation (89), or when (approximately):

$$\dot{p} > \frac{4.868 \times 10^{-2} A P_o (P_o + \Delta P)}{m_o \sqrt{T}} \quad \ldots\ldots\ldots\ldots \quad (92)$$

in the case of equation (90).

These criteria yield the data shown in Table 4 for the two vessels considered in the reference and depicted herein by Photographs 12, 13 and 14. The calculating data are cited below the derived data.

The pressure exerted on a tubular lamp-post, by a blast wave is

$\dfrac{dp}{dr} \dfrac{\pi D}{4}$ where D is the diameter of the lamp-post.

The corresponding total, but uniformly, distributed load (W) on the lamp-post is

$$\frac{dp}{dr} \quad \frac{\pi D^2}{4}$$ h where h is the height of the lamp-post above the point of cantileverage. That is,

$$W = \frac{\pi h D^2}{4} \quad \frac{dp}{dr} \quad \ldots\ldots\ldots\ldots\ldots\ldots\ldots\ldots\ldots \quad (93)$$

For $\frac{dp}{dr}$ may be substituted either of the expressions given in Equations (32) or (33), according to the circumstances.

Table 4 – *Data for damage to vessels at Flixborough*

	Vessel C2525 (Photograph 14)	Vessels R2525 and R2526 (Photographs 12 and 13)
Minimum rate of pressure rise for crushing just to occur	2.016×10^5 Pa s^{-1} (Eq 91) 3.45×10^5 Pa s^{-1} (Eq 92)	8.188×10^6 Pa s^{-1} (Eq 91) 9.384×10^6 Pa s^{-1} (Eq 92)
Time to reach crushing condition at minimum rate of pressure rise	1.3 s (Eq 89) 1.0 s (Eq 90)	0.185 s (Eq 89) 0.18 s (Eq 90)
Crushing pressure at minimum rate of pressure rise	2.621×10^5 Pa (Eq 89) 3.45×10^5 Pa (Eq 90) ($\Delta P_{cr} = 1.311 \times 10^5$ Pa)	1.518×10^6 Pa (Eq 89) 1.689×10^6 Pa (Eq 90) ($\Delta P_{cr} = 7.59 \times 10^5$ Pa)
Rate of pressure rise necessary to attain crushing pressure in:		
0.05 s	2.673×10^6 Pa s^{-1} [1.337×10^5 Pa]	1.755×10^7 Pa s^{-1} [8.775×10^5 Pa]
0.1 s	1.363×10^6 Pa s^{-1} [1.363×10^5 Pa]	1.039×10^7 Pa s^{-1} [1.039×10^6 Pa]
0.2 s	7.101×10^5 Pa s^{-1} [1.420×10^5 Pa]	–
0.4 s	3.873×10^5 Pa s^{-1} [1.549×10^5 Pa]	–
[and the associated atmospheric pressure]		

Calculating Data

A	0.6452 m^2	0.5394 m^2
P_o	101 325 Pa	101 325 Pa
m_o	106.8 kg	12.73 kg
T	373°C	373°C
ΔP_{cr}	1.311×10^5 Pa	7.59×10^5 Pa

Conventional stress analysis states that the deflection of the free end of a cantilever is related to the total load, uniformly distributed, by

$$d = \frac{Wh^3}{8EI} \qquad (94)$$

where, in this context, E is the modulus of elasticity of the material and I is the second moment of area of cross-section about the neutral axis. Therefore:

$$d = \frac{\pi D^2 h^4}{32EI} \quad \frac{\mathrm{d}p}{\mathrm{d}r} \qquad (95)$$

If yield has occurred predominantly over a short section near the root of the contilever (*eg* Photograph 25) then the deflection (d') at the point of transition from deformed to undeformed metal is given approximately by

$$d' = d \frac{\frac{l}{2}}{h - \frac{l}{2}}$$

where l is the length of the deformed section. In this case the loading giving rise to the deflection d' (assumed to be applied distance l from the root of the cantilever) is

$$W \frac{h-l}{h}$$

as a point load at l and $W \frac{l}{h}$ as a distributed load over l. Whence

$$d' = \frac{1}{3} W \frac{(h-l)}{h} \frac{l^3}{EI} + \frac{1}{8} W \frac{l}{h} \frac{l^3}{EI}$$

or

$$d' = \frac{Wl^3}{hEI} \left(\frac{1}{3} (h-l) + \frac{1}{8} l \right)$$

or

$$d' = \frac{Wl^3}{hEI} \left[\frac{h}{3} - \frac{5l}{24} \right] \qquad (97)$$

ie

$$d' = \frac{\pi l^3 D^2}{12EI} \left(h - \frac{5}{8} l \right) \frac{\mathrm{d}p}{\mathrm{d}r} \qquad (98)$$

It is to be emphasised that these equations apply strictly only in the elastic range of deflections.

If it is assumed that deformation takes the form of an arc of a circle then it can be shown (for elastic-reversible deformation) that approximately:

$$d' = \frac{l^2}{4D} \quad \frac{\sigma}{E} \qquad \text{(99)}$$

where σ is the fibre stress in the outermost fibres. Permanent "plastic" deformation occurs when the yield stress in the outermost fibres is achieved. The boundary conditions giving rise to this state are defined by

$$\left(\frac{dp}{dr}\right)_y = \frac{24\sigma_y I}{\pi l\,(8h - 5l)\,D^3} \qquad \text{(100)}$$

where the suffix indicates correspondence with the yield condition.

For the deformation portrayed in Photograph 25 $l = 1.25$ m (approx); $h = 7.47$ m, $D = 0.1143$ m; $\sigma_y = 2.76 \times 10^8$ Pa (SAE 1020 annealed – 0.2% offset) $I = 2.334 \times 10^{-6}$ m^4 and

$$\left(\frac{dp}{dr}\right)_y = 4.93 \times 10^4 \text{ Pa.}$$

The corresponding yield deflection d'_y is 0.0046 m. The actual deflection, from scaling the photograph, at the position of d' was 0.205 m. It would appear, therefore, that the total deflection during the blast at the position l (d') may well have been about 0.21 m. Such a deflection, under the geometric conditions assumed above, would impart elastic deformation to the pipe material within 0.005 m of the neutral plane and either side of it, and permanent deformation to the remaining 0.052 m of material beyond and either side of the region of elastic behaviour.

Assuming that the extension (or compression) of the material of the lamp-post is linearly related to its distance from the neutral plane, and assuming, as is implicit in the whole of the treatment, that the lamp-post cross-section remains circular, then the work done may be calculated. Further simplifying assumptions may be made (which are not essential): that the compressive and tensile stresses are proportional to the extension up to the plane of yield, that these stresses have the same magnitude, and that beyond yield for the remainder of the section the stress is constant (whether compressive or tensile) at the yield value *ie* 2.76×10^8 Pa for this example.

A total deflection, d', of 0.21 m means an extension (or compression) of material of 0.672 m/m from the neutral plane. As the elastic deformation occupies such a small region of material, the total elastic strain work may be adequately approximated by

$$W_e = \frac{0.6022\,E\,b\,x^3}{l} \qquad \text{(101)}$$

where b is the tube wall thickness and x the depth of elastic deformation.

From the lamp-post already considered

$$W_e = 3.5\text{J}$$

This is small and can usually be ignored in comparison with the work of permanent deformation which can be stated approximately by

$$W_p = 0.90\,\sigma_y \left[\frac{2b\left(\frac{D}{2} - b\right)^3}{D - b} + \left[b\,(D - b) \right]^{3/2} \right] \quad \ldots\ldots\ldots\ldots \quad (102)$$

in which the symbols have the meanings set out above. For the lamp-post shown in photograph

$$W_p = 5.65 \times 10^3\,\mathrm{J}$$

This energy was expended during the passage of the leading edge of the blast wave which must have maintained a sufficiently high pressure gradient for sufficient time for the deformation to occur. Of course, higher pressure gradients than that having the minimum value could have done the damage. But pursuing the minimum value, for the sake of argument, at the position of this lamp-post from the "centre" of blast, about 100 m (*ie* virtually at the edge of the cloud), a pressure gradient of 4.93 x 10^4 Pa/m corresponds, equation (33), to the release of at least 1.72 x 10^{12} Joule – equivalent to the rapid combustion of 40 tonnes of cyclohexane.

In this subsection the damage already considered has been of a specialised kind. General equations have not yet been derived or presented. Analogously to explosive pressure waves, "blast" waves will move objects according to their masses, their shapes and their restraints. Let us consider the simplest object first – an unrestrained but initially stationary body. By equating its inertia with the pressure force its equation of motion may be formulated.

$$m\ddot{x} = S_e\,L_e\,\frac{\partial p}{\partial r} \quad \ldots\ldots\ldots\ldots \quad (103)$$

as x and r are equivalent

$$m\ddot{v} = S_e\,L_e\,\frac{\partial p}{\partial r}$$

$$ie \quad \frac{md^2r}{dt^2} = S_e\,L_e\,\frac{\partial p}{\partial r} \quad \ldots\ldots\ldots\ldots \quad (104)$$

Now p is a function of both r and t as indicated by the partial differential. Substitution of an appropriate expression for

$$\frac{\partial r}{\partial t}$$

should provide a solution for movement for all conditions. In fact, probably sufficient information is obtained by assuming that the phases of pressure rise and fall are linear either side of the maximum pressure position, *ie*

$$m\,\frac{\partial^2 r}{\partial t^2} = S_e\,L_e\,\frac{\Delta P}{\Delta L} \quad \ldots\ldots\ldots\ldots \quad (105)$$

which may be integrated to

$$\frac{\partial r}{\partial t} = \frac{S_e L_e}{m} \frac{\Delta P}{\Delta L} t_o \quad \ldots\ldots\ldots\ldots\ldots\ldots \quad (106)$$

and
$$r = \frac{S_e L_e}{2m} \frac{\Delta P}{\Delta L} t_o^2 \quad \ldots\ldots\ldots\ldots\ldots\ldots \quad (107)$$

ΔP is the pressure maximum over atmospheric pressure and ΔL is the half period of the pressure wave. Equations (106) and (107) give values of

$$\frac{\partial r}{\partial t} \text{ and } r$$

for half the wave only. The full movement of an object is $2r$.

Reference to the theoretical work done by the Building Research Station has already been made[149,150]. This, although referred to in other publications[208,209], has not as yet appeared in a comprehensive and coherent form. Mainstone has, nevertheless, extended this treatment recently[210] and clearly the approach is well developed. As mentioned earlier, one of the forms of pressure wave considered by Mainstone (and L.S. Vallance) was that having a sine-wave profile. Accordingly the deductions possible from that theory, for confined explosions, should be appropriate to the UVCE situation if the same pressure wave form develops. Mainstone's approach has been to consider the response of a simple spring-mass system, entirely analogously to Strehlow and Baker[86] for a shock wave.

The "spring" considered in the theory has a linear (elastic) stress/strain response to a yield condition after which the strain becomes plastic – deformation at constant stress (the yield stress) is assumed to continue thereafter. This was the criterion which was applied to the lamp-posts earlier. The sinusoidal pressure wave form was defined by Mainstone by maximum pressure and wave period. The theoretical results have been presented, amongst other ways (*eg* Figure 14 of Mainstone[150]), by plotting the ratio of maximum pressure to yield stress against the ratio of wavelength to characteristic (elastic) vibration period for different values of the parameter: the ratio of maximum (elastic plus plastic) strain to maximum elastic strain.

This theory indicates that for values of the ratio of the wave period to the vibrational period below about 0.5 of the ratio of yield stress to maximum pressure is proportional to the ratio of wave to vibrational periods. For wave period to vibrational period ratios in excess of 10 the ratio of yield stress to maximum pressure tends to unity. Between 0.5 and 10 those spring/mass systems displaying considerable plastic behaviour (*eg* where the ratio of total strain to elastic strain exceeds say 3) respond as overdamped systems in their subsequent motion. For small or zero ranges of plastic behaviour, however, overshooting of the equilibrium condition is predicted with subsequent damped oscillation to the equilibrium condition. The maximum initial overshoot for purely elastic response produces about 50% more stress than the equilibrium stress appropriate to a pressure force of the same magnitude statically applied. This result has been shown by other workers[86,203].

Applying Mainstone's theory to mild steel lamp-posts indicates a possibly critically damped response – given: a sine-form pressure wave; a "typical" maximum ratio of total strain to elastic strain of 1.5 to 2; and a sufficiently large ratio of wave to vibrational period. In other words, mild steel lamp-posts should be capable of providing the most accurate information concerning UVCEs because of their critically damped response and their relatively short vibrational periods.

KEY TO SYMBOLS USED IN THE TEXT

A	The area of an opening or openings in hollow vessels (used in equations 85 – 92 inclusive)
A	Area of glass pane
B	Derived parameter (equation 69)
C	Constant (equation 79)
C_d	Drag coefficient
D	Diameter
D	Damage intensity
E	Energy release
E	Modulus of elasticity (used in equations 94 to 101)
E_o	Total energy release
E_e	Explosive energy release
E_u	Unconfined vapour cloud explosion energy release
F, G	Constants (equation 70)
ΔH	Heat release per unit mass of flammable material
I	Impulse (equation 39)
I	Second moment of area of cross section (equations 94 to 100)

$\bar{I}$	Dimensionless impulse (equation 81)
K	Dimensional constant (used in equations 53 to 57)
K	Constant of elastic proportionality (equations 74 to 81)
K_1, K_2, K_3, K_4, K_5	Dimensionless constants (used in equations 1 to 7)
L	Thickness or length of an object measured in a radial direction from the centre of explosion
L_e	Effective length of object in direction of maximum pressure gradient
ΔL	Length over which pressure difference ΔP operates (equation 105)
M	Molecular weight
M'	Molecular weight of a flammable mixture
M_a	Molecular weight of air
M_o	Molecular weight of material issuing from an orifice
M_v	Molecular weight of vapour
P	Pressure
$\bar{P}$	Dimensionless pressure (equation 80)
ΔP	Pressure difference
P_{cr}	Pressure difference necessary to crush a hollow vessel
P_i	Internal pressure
P_o	External pressure
P_s	Dimensionless pressure (equations 70 & 71)
Q	Rate of evaporation (kg/s)
Q	Dimensionless dynamic pressure (equation 69)

Q_1	Specific rate of evaporation in one minute (kg/m^2)
Q_{Fl}	Quantity of flammable in admixture with air within the flammable limits arising either through momentum jet or buoyant plume mixing
R	Radial distance from centre of explosion
R_o	Radial distance of the furthest edge of a pressure wave at the conclusion of the explosion
R_1 R_2	The nearer and further radial extremities of an object experiencing a pressure wave from an explosion, focussed at the centre
R_m	The mean distance of an object from the centre of explosion measured radially (equal to $\frac{R_1 + R_2}{2}$)
R_s	The radial distance at which a shock wave will appear as a result of a gradual pressure disturbance at the centre
S	Area of object presented to wind (equation 67)
S_e	Effective surface area of object exposed to pressure (equation 103)
T	Absolute temperature
T_o	Temperature of fluid issuing from an orifice
T_a	Absolute temperature of the atmosphere
T_b	Boiling temperature at one atmosphere-degrees absolute
T_L	Initial temperature of liquid
T_s	Initial temperature of substratum
U V	Dimensionless velocities (equation 69)
V	Volume
V_{Fl}	Total volume of mixture of flammable with air arising from momentum jet or buoyant plume mixing from orifice
$\dot{V}_o$	Volumetric rate of emission of flammable material

W	Total load resulting from a distributed pressure force
W	Yield of explosive detonation (equation 73)
W_e	Work done during elastic deformation
W_p	Work done during permanent deformation
$\dot{X}$	Dimensionless velocity (equation 69)
Z	Dimensionless time (equation 69)
a	Velocity of sound in the normal atmosphere
a'	Velocity of sound in a flammable mixture
b	Thickness
c	Isothermal velocity of sound in a two-phase mixture
c_l	Lower limit of concentration of flammable in air
c_L	Specific heat of liquid
c_u	Upper explosive limit of flammable in air
c_s	Specific heat of substrate
d	Deflection
d'	Deflection of permanently distorted material
d_o	Diameter of orifice
f	Acceleration
g	Acceleration due to gravity
h	Height
Δh	Combustion heat release per unit volume
k	Dimensionless constant in the exponent of the expression relating shock pressure to pressure behind shock wave
l	Length (or height) of deformed section

m	Mass of flammable contributing to rapid energy release in a UVCE
m	Mass of gas
m	Mass of rigid object exposed to a pressure force
m_o	Original mass of air within hollow container
$\dot{m}$	Mass rate of flow
p	Pressure
p_1 p_2	Pressure experienced by leading and far edges of an object experiencing a pressure wave
$\dot{p}_c$	Rate of pressure rise at the centre of the explosion
p_w	The pressure within a cone of space occupied by a flammable mixture
p_r	Peak reflected over-pressure
p_o	The pressure in air surrounding a conical sector containing a flammable mixture
p_s	Peak side-on pressure
q	Dynamic wind velocity (equation 68)
r	Radial distance from the centre of blast symmetry
r_b	Radius of the rear edge of a steep fronted shock wave from the centre of blast symmetry
t	Time
Δt	The minimum time for the development of crushing pressures of hollow vessels
t_o	The time of conclusion of rapid energy release in a UVCE
t_l	The characteristic shock loading time
t_s	The time taken for a pressure wave to transform into a shock wave
t_u^+	The duration of the positive phase of the dynamic wind

t_p^+, t_+	The duration of the positive phase of shock over-pressure
v	Flame speed
v_f	Final flame speed
$\bar{v}$	Average flame speed
w	Pressure ratio across an orifice
w_o	Velocity of discharge from an orifice
x	Linear distance
z_l	Axial extent of momentum jet or plume to the point of lower limit of flammability
z_{tr}	Axial distance of the point of transition from momentum jet mixing to plume mixing
α	Cone semiangle
β	Acceleration coefficient (equation 67)
γ	Ratio of specific heats for flammable mixture
δ	Coefficient of exponent (equation 70)
ϵ	Stoichiometric volume ratio
ζ	Coefficient of exponent (equation 70)
η	Efficiency
λ	Latent heat
λ_B	Latent heat of boiling
μ	The mass ratio of gas to liquid in a two-phase system
ν	Coefficient of exponent (equation 71)
ρ	Density
ρ_V	Vapour density

ρ_L	Liquid density
ρ_s	Substrate density
ρ_b	A derived parameter used in equation 35
σ	Stress
σ_y	Yield stress
τ	Time – after the conclusion of the rapid energy release phase of a UVCE
ϕ	Maximum pressure in the propagating pressure wave after the period of rapid energy release has terminated
$\bar{\phi}$	Average maximum pressure of a propagated pressure wave (used in equations 39 and 40)
ω	Natural frequency of oscillation of spring/mass system

UNITS

SI units have been used throughout, supplemented where appropriate by equivalents in other unit systems. The following conversions may be used:

1 psi	=	6900 Pa
1 bar	=	10^5 Pa
1 atmosphere	=	$1.013\ 25 \times 10^5$ Pa
1 Newton/m^2	=	1 Pa
1 Kilogramme-force (kgf) or kilopond (kp)	=	9.807 N
1 kgf/cm^2 (or kp/cm^2)	=	0.9807×10^5 Pa

REFERENCES

1 Private communication

2 Lewis, Bernard and von Elbe, Guenther, "Combustion, Flames and Explosions of Gases", (Academic Press, New York, 1961)

3 Khitrin, "The Physics of Combustion and Explosion", (N.S.F. Translation from Russian by the Israel Prog for Sci Transl, 1962)

4 Gaydon, A.G. and Wolfhard, H.G., "Flames", (Chapman and Hall, London, 1970)

5 Strehlow R.A., "Fundamentals of Combustion", (International Textbook Co, Scranton, Pennsylvania, 1968)

6 Williams, F.A., "Combustion Theory", (Addison-Wesley, Reading, Massachusetts, 1965)

7 Jost, W. "Explosion and Combustion Processes in Gases", (McGraw-Hill, New York, 1946)

8 Jouget, E., "Mechanics des Explosifs", (Dain, Paris, 1917)

9 Spalding, D.B., "Some Fundamentals of Combustion", (Butterworths, London, 1968)

10 Zeldovich, Da. B., "Theory of Flame Propagation", (English Translation, NACA Tech Memo 1282, 1948)

11 Mullins, B.P. and Penner, S.S., "Explosions, Detonations, Flammability and Ignition", AGARG Monograph No 31 (Pergamon Press, New York, 1959)

12 Kanury, A. Murty, "Combustion Phenomena", (Gordon and Breach, New York, 1975)

13 Mecklenburg, J.C., "Plant Layout", (Leonard Hill/Institution of Chemical Engineers, London 1973)

14 Taylor, G.I., *Proc Roy Soc* (1946) **A186**, 273

15 Rasbach, D., private communication

16 Spiegel Schaden, December 1969

17 Bradford, W.J. and Culbertson, T.L. *Loss Prevention,* (1967), **1**, 28

18 Spielvogel, Ing. E., private communication

19 Ball, J.G., *The Chemical Engineer* (April 1976) 275

20 Cox, J.I., *The Chemical Engineer* (May 1976) 353

21 Gugan, K., *The Chemical Engineer* (May 1976) 341

22 Adcock, C.T. and Weldon, J.D., *Loss Prevention* (1967) **1**, 70

23 Stull, D.R., "Fundamentals of Fire and Explosion", AIChE Monograph Series, **73**, 10, (1977)

24 Weinberg, F.J., "Flame Processes", (Imperial College, London, 1960)

25 Coward, H.F. and Jones, G.W., *U.S. Bureau of Mines Bulletin 503,* (1952)

26 Bádami, G.N. and Egerton, A.C., *Proc Roy Soc,* (1955) **A228**, 297

27 Dixon-Lewis, G. and Isles, G.L., "Seventh Symposium on Combustion", 475 (Butterworths, London, 1959)

28 Andrews, G.E., Bradley, D. and Lwakabamka, S.B., "15th Symposium (International) on Combustion", 655, (1974)

29 Oppenheim, A.K. and Soloukhin, R.J., *Annual Review Fluid Mech,* (1973) **5**, 31

30 Gordon, W.E., "Third Symposium on Combustion *etc*," 579 (Williams and Wilkins, Baltimore, Maryland, 1949)

31 Jones, A.R., Schwar, M.J.K. and Weinberg, F.J., *Proc Roy Soc* (1971) **A322**, 119

32 Bull, D.C., Elsworth, J.E., Hooper, G. and Quinn, C.P. *J Phys D* (1976) 9, 1991

33 Bull, D.C., Elsworth, J.E. and Hooper, G., "Sixth International Colloquium on Gasdynamics of Explosions and Reactive Systems", (Stockholm, 1977)

34 Woolfolk, R.W., S.R.I. Final Report for Contract No 0017-69-C-4432, 25 January 1971 (Stanford Research Institute)

35 Ablow, C.M. and Woolfolk, R.W., S.R.I. Final Report for Contract No 0017-71-G-4421, December 1972 (Stanford Research Institute)

36 Woolfolk, R.W. and Ablow, C.M., "Proc Conf Mechanism of Explosion and Blast Waves" **IV** (USA, 1973)

37 Maurer, B., Hess, K., Giesbrecht, H. and Llickel, W., "2nd International Symposium on Loss Prevention", **VI**, 305 (Heidelberg, W. Germany, 1977)

38 Burgoyne, J.H. and Cohen, L., *Proc Roy Soc* (1954) **A.225**, 375

39 Kinney, G.F., "Explosive Shocks in Air" (McMillan, New York, 1962)

40 Bradley, J.H., "Shock Waves in Chemistry and Physics", (Methuen, London, 1962)

41 Baker, W.E., "Explosions in Air" (University of Texas Press, Austin, Texas, 1973)

42 Peters, J., *Annals of the New York Academy of Sciences* (1968) **152**, 283

42a Golde, R.H., "Lightning Protection", (Edward Arnold, London, 1973)

42b Uman, M.A., "Lightning" (McGraw Hill, London, 1969)

43 Strehlow, R.A., "14th Symposium (International) on Combustion" **1189** (The Combustion Institute, Pittsburgh, Pennsylvania, 1973)

44 Davenport, J.A., *Loss Prevention* in the course of publication. (Paper presented to 83rd National Meeting of AIChE, Houston, Texas, March 1977, Paper 24a)

45 James, D.J. and Rowe, R.D. AWRE Report No E3/64, (Harwell, 1964)

46 Burgess, D.S. *et al,* U.S. Bureau of Mines Report of Investigations 7196 (1968)

47 Tolond, J., "Ships in the Sky" (Holt, New York, 1957)

48 Beaubois, H. and Demand, C., "Airships" (MacDonald and Janes, London, 1974)

49 *NFPA Quarterly* (1952-53) **46**, 59

50 *NFPA Quarterly* (1952-53) **46**, 80

51 *NFPA Quarterly* (1952-53) **46**, 68

52 *NFPA Quarterly* (1952-53) **46**, 56

53 *NFPA Quarterly* (1952-53) **46**, 61

54 *NFPA Quarterly* (1949-50) **43**, 101

55 *NFPA Quarterly* (1952-53) **46**, 62

56 Marshall, V.C., *The Chemical Engineer* (August 1977) 573

57 *NFPA Quarterly* (1952-53) **46**, 63

58 *NFPA Quarterly* (1952-53) **46**, 57

59 *NFPA Quarterly* (1952-53) **46**, 61

59a Lewis, D.J., Seminar on UVCEs, Imperial College, London, 18-19 April 1978, to be published by Pergamon Press

60 *NFPA Quarterly* (1949-50) **43**, 173

61 *NFPA Quarterly* (1952-53) **46**, 76

62 *NFPA Quarterly* (1952-53) **46**, 65

63 *NFPA Quarterly* (1952-53) **46**, 59

64 *NFPA Quarterly* (1951-52) **45**, 66

65 *NFPA Quarterly* (1952-53) **46**, 240

66 *NFPA Quarterly* (1961-62) **55**, 200

67 *NFPA Quarterly* (1961-62) **55**, 185

68 *NFPA Quarterly* (1961-62) **55**, 198

69 *NFPA Quarterly* (1961-62) **55**, 186

70 *NFPA Quarterly* (1961-62) **55**, 186

71 *NFPA Quarterly* (1961-62) **55**, 193

72 *NFPA Quarterly* (1961-62) **55**, 187

73 *NFPA Quarterly* (1961-62) **55**, 201

74 *NFPA Quarterly* (1961-62) **55**, 203

75 *NFPA Quarterly* (1961-62) **55**, 199

76 Troyan, J.E. and Le Vine, R.Y., *Loss Prevention* (1968) **2**, 125

77 Walls, W.L., *NFPA Quarterly* (1963-64) **57**, 9

78 ARAMCO Report, *NFPA Quarterly* (1963-64) **57**, 88

79 Reider, R., Otway, H.J. and Knight, H.T., *Pyrodynamics* (1965) **2**, 249

80 Kevil, C.G., *Safety in Air and Ammonia Plants* (1967) **9**, 37

81 Dorsay, J.J., *Safety in Air and Ammonia Plants* (1967) **9**, 80

82 *Fire International* (1966) **12**, 16

83 McCarey, J.J., *Fire Engineering* (August 1966) 103

84 Goforth, C.P., *Fire Journal* (1968) **62(3)**, 1

85 "Report concerning an inquiry into the cause of the explosion on 20th January 1968 at thc premises of Shell Nederland Raffinaderij N.V. in Pernis" (Netherlands Ministry of Social Affairs and Public Health)

86 Strehlow, R.A. and Baker, W.E., *Progress in Energy and Combustion Science* (1976) **2**, 27

87 *Fire Journal* (1969) **63(3)**, 54

88 Kolgar, *Loss Prevention* (1971) **5**, 26

89 Watrous, L.D. *Fire Journal* (1971) **65**, 45

90 *Fire Journal* (1971) **65(2)**, 45

91 Burgess, D.S. and Zabetakis, M.G., U.S. Bureau of Mines Report of Investigations **RI** 7752 (1973)

92 *Fire Journal* (1971) **65(4)**, 50

93 "The Flixborough Disaster – Report of the Court of Inquiry" (HMSO, London, 1975)

94 *Fire Journal* (1975) **69(1)**, 59

95 Proyard, J., private communication

96 Saia, S.A., *Loss Prevention* (1976) **10**, 23

96a Madge, C.H., private communication

97 "Rapport over de Explosie bij D.S.M. te Beek (L) 7 November 1975. Gaswolkexplosie in Naftakranker II" (Nederlands Ministeria van Sociale Zaker)

98 *Fire Command* (1977) **44(a)**, 28

99 Ricou, F.P. and Spalding, D.B., *J Fluid Mech* (1961) **11**, 21

100 Marshall, J.G., "Symposium on Chemical Process Hazards", **VI** (I Chem E Symposium Series No 49, 1977)

101 "Factory Mutual Engineering", *Loss Prevention Bulletin* No 36.10 "Properties of Flammable Liquids, Gases and Solids"

102 Zabetakis, M.G., US Bureau of Mines, Bulletin 627

103 Jarrett, D.E. *Annals of the New York Academy of Sciences* (1968) **152**, 18

104 Burgoyne, J.H., Bett and Lee, "Third Symposium on Chem Proc Haz" (1968)

105 Gugan, K., "Third Symposium on Chem Proc Haz" (1968)

106 Rutledge, P.V., *Chem Ind* (1966) 34

107 Brasie, W.C. and Simpson, D.W., *Loss Prevention* (1968) **2**, 91

108 Ball, J.G., *Process Engineering* (December 1975) 39 and (January 1976) 53

109 Butler, P., *The Engineer* (December 1975) 20

110 Cottrell, A.H. and Swann P.R., *The Chemical Engineer* (April 1976) 266

111 Cox, J.I., *The Chemical Engineer* (June 1975) 342

112 Gugan, K., Institution of Chemical Engineers Meeting, "Flixborough – the Aftermath", Edinburgh, Scotland, June 1975

113 Hughes, J.O., ibid

114 King, R.W., Institution of Chemical Engineers Meeting, "The Technical Lessons of Flixborough", Nottingham, England, December 1975

115 Marshall, V.C., *The Chemical Engineer* (November 1975) 661

116 Marshall, V.C., *Chemical Engineering* (December 1975) 22, 58

117 Munday, G., *The Chemical Engineer* (April 1976) 278

118 Nash, P. and Theobald, C.R., Institution of Chemical Engineers Meeting, "The Technical Lessons of Flixborough", Nottingham, England, December 1975

119 Rakestraw, J.A., *Chemical Age* (September 1975) 6

120 Tucker, D.M., Institution of Chemical Engineers Meeting, "The Technical Lessons of Flixborough", Nottingham, England, December 1975

121 Warner, F.E., *Chemical Engineering Progress* (September 1975) **71**, 77

122 Wood, R., *Process Engineering* (December 1975) 3

123 Jones, T.B. and Spracklen, C.T., Report submitted to Commission of Inquiry into the Flixborough Disaster, 1974

124 Sadee, C., Samuels, D.E. and O'Brien, T.P., *Journal of Occupational Accidents* (1976/1977) **1**, 203

125 Merz and McLellan, Report referred to in Evidence to the Commission of Inquiry into the Flixborough Disaster, 1974

126 Cremer and Warner, Report No 9 to the Commission of Inquiry into the Flixborough Disaster, January 1975

127 Kemp, P.H., Report submitted to Commission of Inquiry into the Flixborough Disaster, 1974

128 Kuhl, A.L., Kamel, M.M. and Oppenheim, A.K., "14th Symposium (International) on Combustion", 1201 (Pittsburg, 1974)

129 Kogarko, S.M., Adushkin, V.V. and Lyamin, A.G. *Nanchno Teknicheskie Problemy Goreniya i Vzryva,* (1965) **2**, 22

130 US Department of Energy, US Coast Guard and American Gas Association Reports on evaporation and dissipation from LNG spillages. Phase II and III reports in preparation

131 Opschoor, Ir. G., Subproject 4, Report 2 RVO-TNO (Rijswijk, Holland, January 1975)

132 Bull, D.C., Elsworth, J.E. and Hooper, G., "Sixth International Colloquium on Gas Dynamics of Explosions" (Stockholm, Sweden, August 1977)

133 Bull, D.C., Elsworth, J.E., Hooper, G and Quin, C.P. *J Phys D* (1976) **9**, 1991

134 Craven, A.D. and Grieg, T.R., Institution of Chemical Engineers Symposium Series, No. 25, (1968), **41**

135 Tanimoto, S., American Institute of Chemical Engineers, 74th National Meeting, (Tulsa, March 1974)

136 "Report on the Experimental Results of Explosions and Fires of Liquid Ethylene Facilities", published by the Safety Information Centre of the Institution of Safety of High Pressure Gas Engineering, Japan. (Translated for the Insurance Technical Bureau, London by M. Sheldon, 1977)

137 Lind, C.D. *Loss Prevention* (1975) **9**, 101

138 Eggleston, L.A., Henera, W.R. and Pish, M.D. *Loss Prevention* (1976) **10**, 31

139 Watts, J.W., *Loss Prevention* (1976) **10**, 48

140 Vincent, G.C. and Howard, W.B., *Loss Prevention* (1976) **10**, 43

141 Vincent, G.C. and Howard, W.B., ibid, 55

142 Hess, K., Hoffman, W. and Stoeckel, A., "First Symposium on Loss Prevention and Safety Promotion in the Process Industries", (1974), 227

143 Maurer, B., Hess, K., Griesbrecht, H. and Leuckel, W., "Second Symposium on Loss Prevention and Safety Promotion in the Process Industries", (1977), 305

144 Marzais, J.J.L., "Safety in H.P. Polyethylene Plants", CEP Tech Man 42-48 (AIChE)

145 Sadee, C., private communication

146 Campbell, I.J. and Pitcher, A.S., *Proc Roy Soc* (1958) **A243**, 534

147 Semenov, N.I. and Kosterin, S.I., *Teploenergatika* (1964) **11**, 46

148 Glasstone, S., (ed) "The Effect of Nuclear Weapons" (USAEC, 1962)

149 Mainstone, R.J., Building Research Establishment CP 24/76

150 Mainstone, R.J., (ed) Symposium: "Buildings and the Hazard of Explosion", Garston, England, October, 1972

151 Coffee, R.D., Raymond, C.L. and Crouch, H.W. "Testing of Materials for Pressure Relief Vents", Eastman Kodak Company, Department of Manufacturing Experiments, 16 May 1950

152 RVO – TNO Report B1-72-6 043-2520 (1972)

153 Pittsburgh Plate Glass Co, Glass Product Recommendations, Structural Technical Service Report No 101, Pittsburgh, Pa (undated)

154 Astbury, N.F. *et al,* British Ceramic Research Association Special Publication 68 (1970)

155 Mainstone, R.J., Building Research Establishment CP 26/71 (1971)

156 Harris, R.J., Marshall, M.R. and Moppett, D.J., Symposium on Chemical Process Hazards VI, Institution of Chemical Engineers Symposium Series No **49**, 83 (1977)

157 Breitwieser, H. and Grimminger, H., *Zentralblatt fur Industriebau* (1973) **9**, 385

158 Bowen, G.I., Albright, R.W., Fletcher, E.R. and White, C.S., Civil Effects Test Operations, USAEC Report No CEX-58.9 (1961)

159 Baker, W.E., Westine, P.S. and Dodge, F.T., "Similarity Methods in Engineering Dynamics: Theory and Practice of Scale Modelling" (Spartan Books, Rochelle Park, N.J., 1973)

160 Fletcher, E.R. and Bowen, I.G., Lovelace Foundation, Technical Progress Report No DASA.1859, Contract No DA-49-146-XZ-373 (Albuquerque, New Mexico, 1966)

161 Reed, J.W., Pape, D.J., Minor, J.E. and De Hart, R.C., *Annals of New York Academy of Sciences* (1968) **152**, 565

162 Nelson, R.W., *Hydrocarbon Processing,* (1977) **56** (8), 103

163 *Annals of New York Academy of Sciences* (1968) 152

164 US Army, Navy and Air Force: "Structures to Resist the Effects of Accidental Explosion", TM5-1300/NAVFACP-397/AFM 88 – 22 (US Government Printing Office, Washington, 1969)

165 White, C.S. *et al* Lovelace Foundation Report WT-1179, *Operation Teapot,* Project 33.1, 1957

166 Pasquill, F., *Proc Roy Soc* (1943) **A182**, 75

167 Pasquill, F., "Atmospheric Diffusion", 2nd Edition (Ellis Horwood, 1974)

168 Sutton, O.G., *Proc Roy Soc* (1934) **A146**, 701

169 Sutton, O.G., "Micrometeorology" (McGraw Hill, 1953)

170 Cox, R.A. and Hoe, D.R. "Second International Symposium on Loss Prevention *etc* in the Process Industries", VI-359 (1977)

171 Long, V.D., "Second Symposium in Process Hazards", Institution of Chemical Engineers, (1963) 6

172 Mackay, D. and Matsugu, R.S., "Evaporation Rates of Liquid Hydrocarbon Spills on Land and Water", *Can J Chem Eng* (1973) 51

173 Fay, J.A., "Unusual Fire Hazard of LNG Tanker Spills", *Combustion, Science and Technology* (1973) 7, 47-9

174 Lind, C.D., "Explosion Hazards associated with Spills of Large Quantities of Hazardous Materials, Phase I", US Coast Guard Report No NTIS AS-A001242 (1974)

175 American Gas Association, Project IS-3-1, LNG Safety Program, Interim Report on Phase II Work, Battelle Columbus Report (July 1974)

176 Shaw, P. and Briscoe, F., "Vapourization of Spills of Hazardous Liquids on Land and Water", SRD R100 (1978)

177 Burgess, D.S., Murphy, J.N. and Zabetakis, M.G., "Hazards of LNG Spillage in Marine Transportation", US Bureau of Mines Report No NTIS AD-705078 (1970)

178 Boyle, G.J. and Kneebone, A., "Laboratory Investigation into the Characteristics of LNG Spills on Water. Evaporation Spreading and Vapour Dispersion, Shell Research Limited". Report of Work done for American Petroleum Institute (1972)

179 Burgess, D.S., Biordi, J. and Murphy, J.N., "Hazards of Spillage of LNG into Water", US Bureau of Mines Report No NTIS AD-754498, 1972

180 Raj, P.P.K. and Kalelkar, A.S., "Assessment Models in Support of the Hazard Assessment Handbook", US Coast Guard Report No CG-D-65-74 (January 1974)

181 Burgess, D.S. and Zabetakis, M.G., "Fire and Explosion Hazards associated with Liquefied Natural Gas", US Bureau of Mines Report of Investigations No 6099 (1962)

182 McQueen, W. Whipp, R.H. and May, W.G., "Spills of LNG on Water – Vaporization and Downward Drift of Combustible Mixtures", Esso Research and Engineering Company Report No EE61E-72, Report of Work done for American Petroleum Institute (1972)

183 Clancey, V. St. J., private communication

184 Lockhart, R.W. and Martinelli, R.C., *Chemical Engineering Progress* (1949) **45**, (1), 39

185 Tangren, R.F., Dodge, C.H. and Seifert, H.S., *Journal of Applied Physics* (1949) **20**, (7), 637

186 Muir, J.F. and Eichhorn, R., *Proc Heat Transfer and Fluid Mechanics Institute* (1963) **183**, 13

187 Chisholm, D., *Proc I Mech E* (1967-68) **182**, (3H), 145

188 Graham, E.J., National Engineering Laboratory Report No 308 (1967)

189 Ethridge, N.H., BRL Memo Report No 1691 (Aberdeen Proving Ground, Maryland, 1965)

190 Johannsohn, G., *International Defence Review* (1976) **6**, 992

191 Geiger, W., *Nuclear Engineering and Design* (1974) **27**, 189

192 Munday, G. Process Industry Hazards Symposium, Institution of Chemical Engineers Symposium Series 47, (1976) 19

193 Oppenheim, A.K., Kuhl, A.L., Kundstrom, E.A. and Kamel, M.M., *Journal of Fluid Mechanics* (1972) 52, (4), 657

194 Clancey, V.J., "Chemical Process Hazards – VI", Institution of Chemical Engineers Symposium Series No 49, (1977) 111

195 Anthony, E.J., *Journal of Hazardous Materials* (1975) **1**, 289

196 Sivashinsky, G.I., *International Journal of Heat and Mass Transfer* (1974) **17**, 1499

197 Foust, A.S., Wenzel, L.A., Clump, C.W., Mans, L. and Anderson L.B., "Principles of Unit Operations" (John Wiley, New York, 1960)

198 Morton, F., Report Submitted to Commission of Inquiry into the Flixborough Disaster (1974)

199 Bowen, I.G., Richmond, D.R. and White, C.S., Lovelace Foundation Technical Progress Report on Contract No DA-49-146XZ-055

200 Fletcher, E.R., Albright, R.W., Goldizen, V.C., Bowen, I.G., Civil Effects Test Operations, USAEC Report No CEX-59.14

201 Brode, H.L., *Journal of Applied Physics* (1955) **26**, 6, 766

202 Fisher, R.B., Krohn, P.L. and Zuckerman, S., Ministry of Home Security Report BPC, 146/WS11 (University of Oxford)

203 Allan, D.S. and Athens, P., *Loss Prevention* (1968) **2**, 103

204 Bowen, G.I., Fletcher, E.R. and Richmond, D.R., Lovelace Foundation Report No DASA-2113, Contract No DA-49-146-XZ-372 (1968)

205 White, C.S., Jones, R.K., Danon, E.G., Fletcher, E.R. and Richmond, D.R., Lovelace Foundation Report No DNA.2738T, Contract No DASA 01-70-C-0075 (1971)

206 Richmond, D.R., Danon, E.G., Fletcher, E.R., Bowen, I.G. and White, C.S. Lovelace Foundation Report No DASA 1860, Contract No DA-49-146-XZ-372 (1966)

207 Munday, G., "Losses from Explosions: A Workbook for the Estimation of Potential Losses from Percussive Unconfined Vapour Cloud Explosions", to be published early in 1979 by the Insurance Technical Bureau

208 Mainstone, R.J., Building Research Establishment Current Paper CP 11/73 (1973)

209 Mainstone, R.J., Building Research Establishment Current Paper CP 64/74 (1974)

210 Mainstone, R.J. (Editor) "Planning and Design of Tall Buildings", Section on Accidental Loading CL-5. In preparation.

GLOSSARY

(A LIST OF TERMS DEFINED AS USED WITHIN THE CONTEXT OF THIS REPORT)

Arrhenius — The name given to the relationship between rate of reaction and temperature which is fundamentally of the same form for *all* chemical reactions

Autocatalytic — A name given to reactions having the facility of accelerating as they proceed towards completion

Blast — The effect caused by an atmospheric pressure disturbance

Blast boundary — A phrase occasionally used to define the edge of the volume occupied by a flammable mixture at the conclusion of combustion

Blow-off — The extinction process which occurs in all "burner" maintained flames when the rate of supply of flammable gas, whether mixed with air or not, is increased beyond a certain level (different for all gases)

Burning velocity — The velocity of propagation of a self-similar flame burning through a flammable gas/air mixture, as measured perpendicularly to the direction of flame movement

Burnt cloud boundary — See "blast boundary"

Combustion — The process of self-sustained reaction at high temperatures (generally with oxygen or air) accompanied by the liberation of heat and light

Combustible — Capable of combustion, but generally reserved for materials which ignite with difficulty and/or burn slowly

Condensed phase — Solid (and liquid) states of matter, a term often used in connection with explosives

Critical temperature That temperature, above which gaseous substances may not be liquified by pressure alone

Deflagration The "normal" process of combustion of mixtures of gas and air by flame as found on all burners and in most explosions. The flame speed is sub-sonic with respect to the combustion reactants and in most cases less than 10 $m\ s^{-1}$

Detonation The "abnormal" process of combustion of mixtures of gas and air in which flame speeds reach sonic velocity in the combustion products and several times the velocity of sound in the combustion reactants. The combustion zone containing the flame exhibits extremes of temperature and pressure and is associated with considerable momentum effects. Detonation is the usual "combustion" process, however, in explosives

Dissociation The process of temporary and reversible molecular breakdown, which occurs when molecules are subjected to high temperature. The process absorbs heat and thus limits the temperatures achieved in, for example, flames

Elastic That behaviour of a system in which the resultant distortion (or strain) is linearly proportional to the force (or stress) imposed

Energy That property of a system which will do work, liberate heat or cause damage depending upon how it is expended

Internal energy – energy contained within any system

Kinetic energy – energy associated with "linear" motion. Reversibly conserved in many systems by periodic transformation into potential energy and *vice versa*

Potential energy – energy in storage as pressure, rotation or elevation. Reversibly conserved in many systems by periodic transformation into kinetic energy and *vice versa*

Strain energy – that energy arising from reversible distortion in an otherwise rigid elastic system

Exothermic Capable of liberating heat upon reaction

Explosion The sudden release of internal energy, accompanied by blast

Explosible Capable of explosion

Explosive Capable of explosion, but more strictly applied to solid or liquid (*ie* condensed phase) materials whose normal behaviour on reaction is to detonate

Flame The process of reaction between a fuel gas or vapour and and oxidiser, usually air, proceeding at high temperature and usually with the emission of visible radiation and heat

Flammable Capable of burning by flame. Synonymous with inflammable. Usually used in respect of easily ignited materials

Flash The mechanism of instantaneous evaporation occurring when pressure is suddenly reduced to below that of a saturated liquid

Gas An element or compound which cannot be condensed isothermally by pressure alone (*eg,* air, methane, hydrogen, *etc*)

Heterogeneous Having discontinuities of phase, concentration or temperature. Usually applied to systems of solid or liquid dispersion in gas

Homogeneous Without discontinuities of phase, concentration or temperature

Homologue One of a series of compounds of similar but progressively extending molecular structure comprising the same elements

Hydrocarbons Compounds of carbon and hydrogen only comprising natural gas, LNG and LPG at one end of the range to polyethylene, bitumen and waxes. Sub-divided into homologuous groups: aliphatics, olefins, aromatics, *etc*

Ignition The process of initiating self-sustaining combustion in a flammable or combustible material

Ignition delay (1) The period of time between commencement of the creation of a flammable cloud and stable flame propagation through it

Ignition delay (2) The period of time between the application of a source of ignition and the development of a stable combustion process (usually of importance only with exclusively thermal ignition sources, including spontaneous ignition)

Ignition source Any physical or chemical event capable of causing ignition, *ie* a hot surface, a flame, a spark, *etc*

Ignition temperature The lowest temperature at which a mixture of gaseous reactants (usually a flammable gas or vapour and oxygen or air) will ignite by purely thermal means. For any particular mixture displaying the lowest ignition temperature this will generally decrease with increasing pressure

Implosion An event giving rise to blast in which the energy released is derived from either (a) the collapse of a volume initially at sub-atmospheric pressure or (b) flame travel in spherical co-ordinates in which the direction of propagation is radially inwards

Isobaric The condition of constant pressure

Isomer A chemical compound having the same elemental composition and molecular weight as one or more others. Each isomer has different chemical and physical properties which are conferred by the different molecular structures

Isothermal The condition of constant temperature

Lean Weak in concentration. In this context, low in the concentrations of flammable gases or vapours in air, but generally not so low that the mixture is below the lower limit of flammability. (Too lean, however, means below the lower limit of flammability)

Momentum jet A forceful discharge into free space from an orifice or opening. The momentum of the discharge induces the surrounding atmosphere into mixture with the jet material such that the flux of momentum through any cross-section of the jet is constant. The criterion for its existence is that the Reynold's number at the orifice must exceed 3000

Olefin A member of the homologous series known as the *olefins.* The series comprises compounds of hydrogen and carbon only, in which in each member of the series one of the linkages between adjacent carbon atoms is unsaturated, giving rise to strain energy and concomitant reactivity

Plastic That behaviour of a system in which the resultant distortion (or strain) and the imposed force (or stress) are non-linear or are unrelated

Reynolds number A dimensionless parameter characterising fluid flow, being a measure of the momentum to viscous forces existing within a fluid. All moving fluids may be compared and contrasted by their Reynolds numbers. Below about 2000

	flow is laminar, or streamline; beyond about 3000 flow is turbulent
Rich	The converse of lean. High in fuel concentration in air, but generally not so high that such a mixture will not support the passage of a flame through it. Over-rich, however, means beyond the upper limit of flammability
Saturated (1)	The description given to the physical state in which vapour and the liquid from which it is derived are the only species present and co-exist in equilibrium (*eg*, steam and boiling water at normal atmospheric pressure). Saturation conditions exist at all pressures, for any particular material, below its characteristic critical temperature
Saturated (2)	The form of molecular structure, particularly applied to organic compounds, in which adjacent carbon atoms are linked by one bond only; it is a strain-free configuration
Stoichiometric	The precise mixture of air (or oxygen) and flammable material in which just sufficient oxygen is present to combine with all the flammable material to produce fully oxidised products. It is a mixture which is unique for each flammable material and can be calculated precisely. Mixtures close to stoichiometric proportions always display the highest flame temperatures and flame speeds
TNT	A conventional high explosive and the explosive material against which it has become customary to measure the performance of all other high explosives. Latterly UVCE explosions have also been compared with TNT
Unsaturated (1)	The description given to a liquid below its boiling point under the prevailing pressure conditions
Unsaturated (2)	The form of molecular structure, particularly applied to organic compounds, in which one or more pairs of adjacent carbon atoms are linked by two or three bonds. Such a configuration *usually* gives rise to a moderately or severely strained structure which displays characteristic reactivity. The notable exception to this rule is the aromatic series of benzene homologues
Vapour	A material which exists in the gaseous phase but which can be condensed at ordinary temperatures by pressure alone
Wave	A mode of energy transport, within or on the surface of a material, capable of reversibly converting kinetic to

potential energy and *vice versa*. All elastic materials propagate waves. Potential energy in a wave transmitted through a medium exists as pressure. Whence:

Shock wave – A pressure "discontinuity" in which the leading edge of the wave is only a few molecules thick. This wave propagates at velocities in excess of sonic in the ambient atmosphere, depending upon the pressure excess over normal, displayed by the wave

Pressure wave – A more general term, including shock waves, in which a pressure distribution exists within the travelling wave, usually of a more gradual form. Its velocity of propagation is usually sonic with respect to the ambient atmosphere

Excess pressure (or overpressure) – The peak pressure excess over the ambient pressure. This is usually independent of orientation of the observer for gradual pressure waves; for shock waves, however, pressure has two components – "static" and "wind" – and has different magnitudes depending upon the angle of incidence of the wave upon the "receiver"